Advanced
Marine Electrics
and Electronics
Troubleshooting

Advanced Marine Electrics and Electronics Troubleshooting

A Manual for Boatowners and Marine Technicians

ED SHERMAN

E P B M

ECHO POINT BOOKS & MEDIA, LLC

Published by Echo Point Books & Media
Brattleboro, Vermont
www.EchoPointBooks.com

Advanced Marine Electrics and Electronics Troubleshooting
ISBN: 978-1-62654-328-7 (casebound)

Cover design by Justine McFarland

*To the hundreds of
ABYC Certified Marine
Electricians I've had the
pleasure of meeting and
working with over the years.
Keep up the good work!*

CONTENTS

FOREWORD

In 1996 I had the good sense to enroll in a course on diesel maintenance at New England Institute of Technology; that much I can take credit for. What I can't take credit for was my exquisite good luck at landing in a class taught by Ed Sherman. It only slowly dawned on me that I was in the presence of one of the country's great teachers and communicators of marine technology. What I knew immediately, though, was that I was getting my money's worth. In the intervening years Ed has moved on to the American Boat and Yacht Council, where he teaches the teachers on boat technology, and for nearly a decade now, he's brought his incomparable teaching into *Cruising World*'s pages to the benefit of us all.

You hold in your hand a fine example of his teaching. As our boats have grown more complex, truly competent skippers are part systems operators as well as navigators and boat handlers. Consequently, the line between recreational boaters and technicians has blurred. *Advanced Marine Electrics and Electronics Troubleshooting* may aim at the professional technician—and it hits that target squarely—but for the dedicated boatowner who really wants to be self-sufficient, this book provides the guidance to solve the most vexing problems in electrical and electronic systems. If you're like me and you're serious about wanting to understand all the details that make up today's boats, you'll find more than ample payback for every moment spent in Ed Sherman's company. You'll certainly come away smarter.

—TIM MURPHY
Executive Editor, *Cruising World*

PREFACE

Unlike the first two books I've published that discuss marine electrical installations—*The Powerboater's Guide to Electrical Systems* and the revision work for the second edition of the classic *12-Volt Bible* by Miner Brotherton—*Advanced Marine Electrics and Electronics Troubleshooting* is intended to help serve the needs of marine electrical technicians and experienced boatowners who want to use the latest technology and techniques to troubleshoot onboard electrical problems. These earlier works were intended for the less experienced boater; this one is not. You won't see information in this book that explains Ohm's law or how circuits are designed. Rather I've assumed you've already grasped those concepts and have more than a rudimentary understanding of how electrical power, both DC and AC, is distributed around a boat.

Over the years, many have accused me of being somewhat of a "gadget freak" because I'm always conducting seminars or writing magazine articles that discuss the use of sophisticated "toys" of the sort that are mentioned in this book.

It's true that I am continually employing new and often unrecognized equipment in my work—not because I feel any particular need to always have the latest gadgets at hand, but rather because I'm always looking for easier, more efficient, and more exact ways to accomplish the everyday tasks of the marine electrician. In addition, I'm acutely aware that modern boats are not getting any easier to work on—at least not in regard to their electrical systems. As the level of sophistication of the modern boat's electrical system continues to increase, and the convergence of traditionally distinct electrical and electronic systems continues, many of the techniques found within this book are going to become as mainstream as checking the electrolyte level in a battery cell used to be.

I'd like to point out that most of the equipment discussed in this book is really not all that new, and many readers who have perhaps migrated into boating from information technology (IT) or telecommunications may recognize some of this equipment and feel right at home. The new thing here is my application of this equipment to the marine environment and the implications of the findings generated by the equipment.

As boat electrical and electronic systems continue to evolve and to look more like the computer network of a modern business office, the techniques described in *Advanced Marine Electrics and Electronics Troubleshooting* will increasingly become the best—and in some cases, the only—way to get the job done. The digital volt-ohm meter is not dead, of course, but if it's the only tool at your disposal for electrical troubleshooting, you simply will not be able to compete as a professional marine technician. And if you're not a technician, but rather the owner of an electronically sophisticated boat, you'll find techniques here that will make you the master of your boat, not its slave.

If you are looking for ways to save money on equipment for electrical troubleshooting, or want to learn how to diagnose electrical problems with a traditional, low-cost 12-volt test light, this book is not for you. But if you want to see how the pros are beginning to do things, read on. The equipment available to the marine troubleshooter keeps getting better and more sophisticated as more and more microprocessor capabilities are integrated into new meters. The value that you derive from the use of this equipment will quickly outweigh the initial purchase price, and the benefits will continue to accrue as time goes on. I hope you benefit from the information and approaches I've presented in this book and, as always, happy boating.

ACKNOWLEDGMENTS

Besides the many technicians I've had the pleasure of working with, this book would not have been possible without the help and guidance of several individuals and two organizations in particular.

First, I'd like to thank the American Boat and Yacht Council in general, and most especially for the technical guidance provided by its *Standards and Technical Reports for Small Craft*. Without these standards we (boaters, field technicians, and I) would be flying blind when performing electrical installations.

Next I'd like to thank the National Marine Electronics Association for its *Installation Standards for Marine Electronic Equipment Used on Moderate-Sized Vessels*. This document is the guiding light for the proper installation of the electronic equipment discussed in the book.

I'd especially like to express gratitude to three individuals for their peer review work and suggestions. Mickey Smith of Boat Systems in Stuart, Florida, is one of the most innovative electrical systems designers working today, and has provided me with much insight into today's advanced marine electrical systems, and what is just around the corner for tomorrow. Jim Vander Hey of Cay Electronics, Portsmouth, Rhode Island, is one of my former students, and has advanced to the leading edge of electronics installation, and currently serves as a member of the board of directors for the NMEA. I'm sure he has no idea how proud it makes me to see his successes. Finally, I'd like to thank Dan Cox of Midtronics, who got me interested in battery conductance testing methodologies and, most recently, in the applications of this technology for troubleshooting voltage drop problems.

Advanced
Marine Electrics
and Electronics
Troubleshooting

Marine Troubleshooting the Modern Way

*A*dvanced Marine Electrics and Electronics Troubleshooting introduces the technician or boatowner to some new equipment for troubleshooting electrical systems. With this equipment, some of the traditional techniques used by marine electricians over the years can now be altered or vastly improved upon. The new gear provides more accurate information, and many of the traditional troubleshooting steps can be eliminated altogether. The microprocessors in these new devices simplify diagnoses that used to require multiple steps and painstaking labor.

Because much of the information in this book may be new to many readers, it seems prudent that we first look at some common problems related to electrical and electronic systems, and then move on to finding solutions to these problems. Rather than using the traditional tools of electrical troubleshooting, however, I will identify the new equipment and point you to the chapter that describes how the equipment is used.

My goal throughout this book is to elevate your skills to the next level in diagnostic work, using the latest in equipment and techniques. If you're a professional electrician, you will certainly need these skills as boats continue to evolve into increasingly more complex electrical/electronic marvels. And if you're the owner of a sophisticated boat, learning to use this equipment will greatly simplify your life when it comes to diagnosing a problem quickly and with minimal electrical expertise. The equipment really does most of the detective work for you. Your part consists of understanding and applying the results generated by the gear, and I hope this book gives you much greater insight into that.

With only a few exceptions, all of the instruments discussed cost less than $500—some of them are less than $100 (and one is under $10!). For anyone who works on boats with complex electrical systems and lots of electronics, these costs can be easily and quickly justified by savings in time and increased effectiveness over old troubleshooting methods. As for the more expensive gear, even that can generate relatively quick payback for professionals, while some advanced amateurs who readily drop thousands of dollars on boat "toys" should have no trouble finding practical applications for these useful and highly capable tools.

To help you build your high-tech toolbox, I've placed Toolbox sidebars in every chapter describing an instrument. In them, I've listed some models and brands that have a good track record to give you a good place to start.

Throughout this book, I've included extensive descriptions of specific instruments to demonstrate the functionalities and basic operations for a *class* of equipment. My use of these instruments is not an endorsement for any particular brand. Instruments from other manufacturers may offer comparable or superior functionality to the ones I've used here. If you are considering the purchase of a specific piece of equipment, carefully weigh all factors: pricing, service, warranty terms, availability, and your specific needs for an instrument's functions and capabilities.

Caution: AC circuitry is dangerous! Extreme shock hazards exist. If you are not experienced and comfortable around AC equipment and troubleshooting, call in a specialist.

FINDING YOUR WAY

How is this book set up? I've divided it into two parts, each with specific goals.

Part 1, Electrical Systems Troubleshooting:

- Identifies new methods for troubleshooting electrical circuits (including the engine starting and battery charging systems and the AC and DC distribution systems).

- Introduces you to test instruments you may not have seen before, such as the time domain reflectometer, megohmmeter, and amp clamp, and explains their capabilities and applications.

- Provides step-by-step instructions, with accompanying photos, on how to use these instruments and apply the results.

- Explains the importance of grounding systems from the multiple perspectives of equipment functionality, shock prevention, corrosion minimization, and lightning protection.

Part 2, Marine Electronics Installation and Troubleshooting:

- Addresses the relationship between the electrical system and marine electronics.

- Looks at numerous installation factors that influence the functionality of electronic equipment, and tells you how to anticipate and avoid problems before you cut holes and run wires.

- Gives special attention to antennas and coaxial cable, as they are critical to the proper functioning of many types of equipment.

- Introduces the cutting edge of onboard wiring—networked systems—to give you a taste of what to expect in the near future.

Let's begin, then, with a discussion of the types of electrical circuits on modern boats, and how to troubleshoot the modern way.

ELECTRICS VERSUS ELECTRONICS

Today's typical modern boat, whether power or sail, has the potential for carrying a wide variety of circuits and equipment:

- AC-supplied battery chargers
- anchor windlass
- audio system
- bilge pumps and blowers
- bow or stern thrusters
- electric galley (stove and oven)
- electric head sanitation system
- electric winches
- engine-driven charging circuit
- engine instrumentation
- engine starter circuit
- hot-water heater
- inverters
- lighting, including navigation lights and general illumination
- navigation and communications electronics
- television

- trim tabs
- refrigeration and air-conditioning system
- shore-power transformers (isolation transformers) and AC generators

All of these devices can be divided into two broad categories: *electrical systems*, which include such things as basic shore-power service or the navigation light circuit on your boat; and *electronics*, which we can generally categorize as circuits that distribute data, such as between a GPS receiver and an autopilot.

The common characteristics of these devices are that they all have a power source, either AC or DC, as well as a path for the electrical current to flow. And inevitably, at some point during their service life, they will need to be repaired, upgraded, or replaced.

Beyond these, few factors apply across all of the devices, so it is convenient to organize them into the categories listed below:

1 Electrical systems
- high-amperage motor circuits
- low-amperage motor circuits
- engine-driven alternator
- lighting circuits
- engine instrumentation
- AC resistive loads (heating elements)
- battery chargers and inverters
- AC power sources

2 Electronics
- communications equipment
- navigation aids
- entertainment equipment

Using these categories, I've created flowcharts to identify some common problems, the equipment or instrument to use to troubleshoot those problems, and the general step-by-step testing procedures. (Each test instrument walks you through the process via on-screen prompts, so I won't repeat all those steps here.) I've also included, for comparison, the old method of troubleshooting, so you can see the greater efficiency and effectiveness of the new method. Finally, each flowchart gives you a quick reference to the chapter that offers more detailed information on the instrument's use and application.

Flowcharts: Electrical Systems

High-Amperage Motor Circuits

The first category consists of motor circuits with high current draws. These are really dual circuits, with a low-amperage control circuit side to activate a relay or solenoid, which then activates the high-amperage side to run the motor. Examples include:

- starter motors
- anchor windlass
- thrusters
- electric winches

The troubleshooting procedure is summarized in the flowchart on page 4.

Low-Amperage Motor Circuits

This category includes AC and DC motor circuits of low current draw (relative to the high-amperage motors above):

- refrigeration systems (AC or DC)
- bilge pumps
- bilge blowers
- trim tabs
- electric heads and macerator pumps

Unlike the circuits for high-amperage motors, low-amperage motor circuits typically do not have a control side and a high-current side, and they are usually fed directly by a DC power supply (although refrigeration systems may be powered by AC or DC, or both).

High-Amperage Motor Circuits

NEW METHOD

Confirm integrity of the power source using a conductance battery tester (Chapter 2).

Test both the control side of the circuit (from the activation switch to the relay or solenoid that activates the high-current side of the circuit) and the high-current side, with or without available power.

If the power source is OK, measure voltage drop using a diagnostic meter (Chapter 2).

If the power source is not OK, recharge or replace the batteries.

SYMPTOM

Slow motor or no motor operation.

OLD METHOD

Check power source integrity with a multimeter, followed by a carbon pile load test of the battery's ultimate condition (not possible with sealed batteries). If OK, use a multimeter to trace the entire circuit, separately measuring each leg of the circuit, recording the voltage readings, and calculating acceptable voltage drop.

Typical problems with this category of circuits include the following:

- No operation when the switch is activated.
- Abnormally slow motor operation.

- Frequent tripping of an overcurrent protection device (such as a circuit breaker or fuse).

The troubleshooting procedure is summarized in the flowcharts on pages 4 and 5.

Low-Amperage Motor Circuits (1)

NEW METHOD

Confirm the integrity of the power source using the conductance battery tester.

The problem may be a tripped breaker or fuse. Remember these devices trip for a reason; simply replacing a fuse may not solve the problem.

Check current draw with an amp clamp and compare to the fuse or breaker rating. Also consider a locked rotor condition. Review Chapter 11 to learn more.

SYMPTOM

No operation when the switch is activated.

OLD METHOD

To determine current draw, connect an ammeter in series; in many cases with higher-current-draw circuits, a resistive shunt is required. Making these temporary, hard-wired connections often takes considerable time.

Low-Amperage Motor Circuits (2)

NEW METHOD

SYMPTOM

Slow motor operation.

Assuming no mechanical problem (such as a seized armature bearing) or extreme solids situation (as in the case of a macerator pump), this is probably a voltage drop issue. A diagnostic meter will pinpoint this in one easy step (Chapter 2). An infrared heat gun (Chapter 5) will also be helpful here.

OLD METHOD

Perform point-by-point voltage drop tests with a multimeter.

The Engine-Driven Alternator

The engine-driven alternator is a unique device—both in its electrical properties and in its function—so it is treated separately.

We'll assume the mechanical side of the alternator is in good order, the drive belts are tight, and the alternator is securely mounted to the engine. The flowcharts also assume that the problem is new, and that previously all was well with the charging system. There are several symptom sets related to alternators that need to be considered:

1 Undercharging will be manifested as poor battery performance.

2 Overcharging will show up as either a foul odor in the battery compartment, or swelling of the battery cases or low electrolyte levels (if the problem is of long standing). With sealed batteries, of course, this cannot be checked, nor can you replenish levels. If a cell of a sealed battery gasses itself dry, the situation becomes dangerous. The cell can short itself out, and an arc inside the battery can cause an explosion.

3 Electronic "noise" emitted from the alternator may interfere with the performance of other electronic equipment on board the boat.

The troubleshooting procedure is summarized in the flowcharts on page 6.

Lighting Circuits

This category includes all lighting circuits, both AC and DC, with their different power supply considerations:

- cabin lights
- navigation lights
- convenience lights

Light circuits are simple. All you need is power supplied from a source and a good return path to ground. The primary concerns are of course electrical continuity and sufficient voltage throughout the circuit, especially in the case of navigation lights, which are governed by both U.S. Coast Guard regulations and American Boat and Yacht Council (ABYC) standards. The equipment can be very basic (e.g., a multimeter) or extremely sophisticated (e.g., a time domain reflectometer, or TDR). I now use a TDR almost exclusively for this sort of work.

The troubleshooting procedure is summarized in the flowcharts on page 7.

Engine-Driven Alternator (1)

NEW METHOD

SYMPTOM

Undercharging or overcharging.

Determine battery condition with a conductance battery tester or diagnostic meter (Chapter 2). This may require connecting the batteries to another charge source to get them to a state of charge that will allow a valid test.

OLD METHOD

Determine open-circuit voltage across the battery posts to determine state of charge. If low, charge battery to at least 70% state of charge, then perform a carbon pile load test (not feasible with sealed batteries). If you don't have access to a carbon pile load tester, disarm the engine so it can't start, and test voltage across the terminals while cranking the engine with the starter motor (this is not as precise as a carbon pile test, but it will indicate an extremely weak battery). Use a multimeter to test voltage output; use an ammeter to test current output.

If the battery is serviceable, look for a voltage differential from no-run to engine running at idle.

If no differential exists, then the alternator is not producing. But why?

Verify that field excitation voltage is being supplied to the alternator: measure the magnetic field strength at the alternator with the engine running versus not running. Use a gauss meter for this step (Chapter 9). *No measurable field* indicates either a fault in the supply circuit to the field or a faulty voltage regulator. *A measurable magnetic field* and *no voltage differential* indicate a fault within the alternator.

The voltage differential between no-run and run should be between 0.5 to 2.5 volts (V). *More than 2.5 V* indicates an overcharge and a faulty voltage regulator. *Less than 0.5 V* (with the alternator running and a lot of DC loads turned on) may indicate an underrated alternator, in which case an upgrade is indicated. To determine if alternator output is enough to meet the demand, perform a DC load analysis by adding the amperage draws for the various appliances.

Engine Instrumentation

Here I've grouped the engine instrumentation and related gauges as follows:

- voltmeter
- tachometer
- oil pressure gauge
- fuel gauge
- temperature gauge

These devices need battery power to function, but they often work on a variable ground principle from a sending unit, or in the case of tachometers, also receive a signal from the

Engine-Driven Alternator (2)

NEW METHOD

SYMPTOM

Electronic noise heard in radio equipment or indicated by a bad signal-to-noise ratio on a Loran-C system.

With the alternator running, test for noise emissions using a transistor radio set off scale on both the AM and FM bands to "home in" on noise (Chapter 9). The frequency of the noise will change proportionally with engine rpm.

Correcting this may require overhauling the alternator or adding a capacitor-type filter on the alternator output.

OLD METHOD

Use process of elimination: shut off one piece of equipment at a time to determine the source of noise.

Lighting Circuits (1)

NEW METHOD

SYMPTOM

No light or abnormally dim light.

If no light, confirm available power and, of course, check the bulb. The TDR works quite well for checking the entire circuit, including the bulb, in one quick step (Chapter 3).

If the bulb is OK, the TDR will indicate a short circuit; if there is a broken filament, it will show an open circuit.

OLD METHOD

Use a multimeter to check continuity through each leg of the circuit.

Lighting Circuits (2)

NEW METHOD

SYMPTOM

Bulb is glowing but abnormally dim.

Voltage drop is the cause. Use a diagnostic meter (Chapter 2) to isolate the location of the drop in one step. The infrared heat gun may show a hot spot in the circuitry, which can cause excessive voltage drop (Chapter 5).

OLD METHOD

Use a multimeter to check continuity through each leg of the circuit.

engine. So in effect, they have two sides to their circuits: the instrument power side, and the sending unit input side.

The troubleshooting procedure is summarized in the flowcharts on pages 7 and 8.

AC Resistive Loads

This category covers AC-powered resistive loads; i.e., heating elements. These include:

- water heaters
- electric ranges and ovens
- electric space heaters

These devices have high current draws and can develop problems over time. You can save a great deal of troubleshooting time by using the diagnostic equipment discussed in this book.

The troubleshooting procedure is summarized in the flowcharts on pages 8 and 9.

Battery Chargers and Inverters

AC-supplied battery chargers, inverters, and inverter chargers have both AC and DC circuitry in the same case. They also have both AC and DC outputs or inputs. The test equipment required will be dictated by whether the problem is on the DC or AC side.

The troubleshooting procedure is summarized in the flowcharts on pages 9–11.

Engine Instrumentation (1)

NEW METHOD

SYMPTOM

Instrument inoperative.

Confirm power and a good ground at the instrument with a basic multimeter.

If no power, repair as needed.

If power and ground are OK, check continuity of the cable between the instrument head and the sending unit. A multimeter is the best tool for this test.

Once the above has been confirmed, check the sender. (Sending units can be checked with a digital volt-ohmmeter [DVOM], but since values are rarely provided in manuals, the results can be misleading.) If grounding the sending unit wire makes the gauge go to full scale with the power turned on, the sender is faulty.

OLD METHOD

The methods described here follow traditional procedures. Sometimes the good old DVOM is still the best way!

Engine Instrumentation (2)

NEW METHOD

SYMPTOM
Instrument reads inaccurately.

In the case of a tachometer, the calibration or cylinder number selector on the gauge may be set incorrectly. The number the gauge is set to should be the same as the number of cylinders in the engine.

In the case of oil pressure, fuel, and other level gauges and temperature senders, the sending unit could be faulty. Try a new sender or check resistance values against a known specification provided by the instrument manufacturer as in the previous flowchart (if you are able to find that information).

OLD METHOD

The methods described here follow traditional procedures. Sometimes the good old DVOM is still the best way!

AC Power Sources

AC power sources consist of shore-power isolation transformers and AC generators, both of which are AC current producers. The transformer is a point of distribution on board the boat; it is supplied by the shore-power feed but electrically isolated from it. Because the tools of choice here are the full arsenal of AC diagnostic tools already mentioned, no flowchart is presented.

- Check amperage and voltage outputs: use the SureTest tool (Chapter 4).

AC Resistive Loads (1)

NEW METHOD

SYMPTOM
Poor performance (power supply suspect).

Use the SureTest tool (Chapter 4) to completely diagnose all power and grounding faults. This tool will save you many steps over using conventional diagnostic methods, and it will perform tests that no other tool can perform.

OLD METHOD

Use a multimeter to confirm presence of AC voltage. Then isolate each cable and test for resistance with an ohmmeter. There is no safe way to check cabling under load, and no way to test trip rates and times.

AC Resistive Loads (2)

NEW METHOD

SYMPTOM
Suspected AC leakage.

Perform safety check for ground fault leakage.

Use a high-resolution clamp-on AC leak tester to isolate the source of AC leakage current by process of elimination (Chapter 10).

OLD METHOD

Use a "split cord" to open the grounding conductor, and check current flow with a high-resolution ammeter in series with the ground wire. Takes more steps and has more potential dangers than a high-resolution leak tester.

AC Resistive Loads (3)

NEW METHOD

SYMPTOM
Excessively high electric bill.

Conduct power consumption analysis.

Use the WattsUp? tool to analyze the equipment's power usage over time (Chapter 6).

Use the WattsUp? tool to calculate operating costs of your equipment.

OLD METHOD

No prior methods existed for performing this task.

AC Resistive Loads (4)

NEW METHOD

Perform a waveform analysis using an oscilloscope (Chapter 7).

SYMPTOM

Equipment operates fine on shore power but malfunctions when powered by an inverter or generator.

OLD METHOD

Without an oscilloscope, there is no efficient way to perform this task.

- Diagnose waveform and harmonics issues: use an oscilloscope (Chapter 7); if a problem is intermittent, use an oscilloscope/laptop combination to track and record data (Chapter 8).
- For safety reasons, routinely use a Yokogawa or similar clamp-on leak tester (Chapter 10).

Flowcharts: Electronic Devices

Here we'll cover circuits serving electronic equipment, such as audiovisual, navigation, and communications devices:

- autopilots
- communications equipment—VHF and SSB radios
- computer or laptop
- depth sounders
- fishfinders and chartplotters
- navigational equipment—GPS and radar
- stereo systems
- televisions and DVD players
- weather monitoring equipment

This category is a bit more complicated than electrical systems because it includes both AC- and DC-powered gear, with slightly different considerations for each. As with all devices, the first step is to confirm the integrity of the power source at the device. In the case of these circuits, that means voltage, whether it's 120 VAC (volts of alternating current) or 12 or 24 VDC (volts of direct current). In many cases, however, you may need to take additional steps.

The troubleshooting procedure for navigation and communications equipment is summarized in the flowcharts on pages 11 and 12. For troubleshooting audio and video systems powered by AC current, see the bottom flowchart on page 12.

Battery Chargers and Inverters (1)

NEW METHOD

Inverter AC output*: Perform a waveform analysis to look for harmonic distortion that may affect the equipment. Use a true RMS multimeter** or, preferably, an oscilloscope (Chapter 7).

SYMPTOM

Equipment operates fine on shore power but malfunctions when powered by an inverter.

*Ideal Industries does not recommend using the SureTest (Chapter 4) on inverters, but I have used it successfully on true-sine-wave inverters.
**True RMS multimeters are discussed in the Powerboater's Guide to Electrical Systems. Simply put, true RMS describes the algorithm a meter must have to accurately analyze a less than perfect AC waveform. Most common meters use an average responding algorithm.

OLD METHOD

Using a conventional multimeter, perform multiple steps using a "split" shore cord adapter to take voltage and current measurements in the hot, neutral, and ground conductors. This method doesn't have the advantages of data logging.

Battery Chargers and Inverters (2)

NEW METHOD

SYMPTOM

Low AC output voltage.

Inverter DC input: Test amperage draw and excessive voltage drop from the battery supply with a diagnostic meter (Chapter 2).

OLD METHOD

Using a conventional multimeter, perform multiple steps using a "split" shore cord adapter to take voltage and current measurements in the hot, neutral, and ground conductors. This method doesn't have the advantages of data logging.

Battery Chargers and Inverters (3)

NEW METHOD

SYMPTOM

Batteries keep going dead. The charger is functioning, and the batteries have been tested and should be able to accept a charge.

Battery charger DC output: An oscilloscope (Chapter 7), or other DVOM with data-logging capability, perhaps connected to a laptop computer if an intermittent problem is suspected (Chapter 8), would be a great choice for tracking both amperage and voltage outputs over time.

OLD METHOD

Using a conventional multimeter, perform multiple steps using a "split" shore cord adapter to take voltage and current measurements in the hot, neutral, and ground conductors. This method doesn't have the advantages of data logging.

Battery Chargers and Inverters (4)

NEW METHOD

SYMPTOM

No output from the battery charger.

Battery charger AC input: Use the SureTest tool to use to confirm proper power supply to the battery charger (Chapter 4).

OLD METHOD

Using a conventional multimeter, perform multiple steps using a "split" shore cord adapter to take voltage and current measurements in the hot, neutral, and ground conductors. This method doesn't have the advantages of data logging.

Battery Chargers and Inverters (5)

NEW METHOD

SYMPTOM

Suspected AC leakage from battery charger or inverter output.

Use the Yokogawa or similar clamp-on leak tester to isolate any ground fault leakage on the AC side of these devices (Chapter 10).

OLD METHOD

Using a conventional multimeter, perform multiple steps using a "split" shore cord adapter to take voltage and current measurements in the hot, neutral, and ground conductors. This method doesn't have the advantages of data logging.

Electronics (1)

NEW METHOD

SYMPTOM

Audible or visual "noise" (the latter seen on readouts or monitors as hash marks or inexplicable lines on the screen).

Confirm proper power to electronic equipment. Use a SureTest tool for AC equipment (Chapter 4) and a DVOM for DC equipment.

Check the integrity of the ground and all of its connections. Use a TDR (Chapter 3) or a diagnostic meter (Chapter 2).

If the gear relies on an antenna, use a TDR to test coaxial connections and cable run. Also, make sure the run is routed correctly as described in Chapter 16.

OLD METHOD

Use a multimeter and/or a trouble light to track through circuits from one termination point to another. Some termination points may be inaccessible.

BASIC TROUBLESHOOTING STEPS

Over the years, I've taught troubleshooting in dozens of classrooms and seminars. While the equipment necessary to solve more complicated problems has become more sophisticated, the basic processes haven't changed. Problem solving is not always easy, and to be done properly and efficiently, several steps must be taken in logical order:

1 **Verify the problem.** Does the complaint represent a genuine problem? Or is the solution really awareness or education?

2 **Gather information.** Get the complaint directly, not thirdhand. Determine the exact symptoms and the conditions under which the problem occurs. Get a timeline—is this a new problem, or has it been going on for awhile? Get as much history about the boat as possible.

Electronics (2)

NEW METHOD

SYMPTOM

One electronic device is not "talking" to another (e.g., a chartplotter or autopilot is not receiving data from a GPS).

Use a TDR to check the integrity of the network cable (Chapter 3).

Use process of elimination if necessary and be sure to check all individual conductors in the cable.

OLD METHOD

Use a multimeter set to ohms to track through circuits from one termination point to another. Some termination points may be inaccessible.

Electronics (3)

NEW METHOD

SYMPTOM

Audio and video systems affected by "bad" power: random diagonal lines showing on screens and background noise is heard over speakers/radio.

Use an oscilloscope to check AC power quality and harmonics (Chapter 7).

If the symptom is only present when the device is powered by one AC power source versus another, the problem is almost always related to AC waveform or harmonic distortion. Use an oscilloscope to pinpoint.

OLD METHOD

Use process of elimination, which is often time consuming and frustrating.

Try to determine the order in which symptoms developed.

3 **Determine probable causes.** Perform a thorough visual inspection, and repair obvious problems before diving into lengthy testing procedures. Have the appropriate service manual or wiring diagrams available if at all possible (unfortunately, this is often not possible). Think about subsystems and ancillary components that could contribute to or cause the problem.

4 **Narrow the list of causes.** Use the history and symptom conditions to narrow down the list of probable causes. If more than one symptom exists, are there common causes? Avoid preconceived ideas. Follow test procedures step by step, skipping none.

5 **Test all subsystems.** Test the most likely cause first. Follow any manufacturer-recommended procedures.

Following these logical steps to troubleshooting may seem like the medical history a doctor gathers when trying to figure out what ails you, and for good reason: the method works. Used in combination with the best, most capable equipment, this method will definitely increase your odds for success. Now let's get started!

ELECTRICAL SYSTEMS TROUBLESHOOTING

When I first started cruising, we got by with oil lamps to read by and no AC shore-power system at all. The DC system provided power for some cabin lights, the navigation lights, engine starting, and an AM/FM radio (with a cassette player), and a basic electronics package consisting of a Loran-C, depth sounder, speed log, and VHF radio. None of this equipment was interfaced, so an onboard network wasn't even a dream at the time. But we got by just fine with that equipment and a good selection of nautical charts, dividers and parallel rulers, and a watch.

Those days are gone, and so too are boats set up so simply. Today, boat buyers' expectations are driving more and more complexity into the installed systems. Along with this trend the need has risen to develop new ways to seek out and troubleshoot problems. Why? Because boatyard labor rates have also increased to the point where, as of this writing, $100-per-hour labor rates for yard services are the norm, and in some places the rates are even higher. Delivering the maximum value for these labor rates is a basic expectation among consumers. They expect things to be done right the first time, and in the shortest amount of time possible.

So in Part 1, we'll look at a variety of gear designed to speed up the diagnostic process and to effectively diagnose equipment that can't be tested using traditional test procedures (such as sealed batteries).

Testing Batteries, Charging Systems, and Starter Circuits, and Measuring Voltage Drop

Of all the areas marine electricians or boatowners must deal with, batteries, alternators, and starter motors, along with their related circuitry, are among the most critical. Boats rely heavily on the integrity of their DC electrical system; on most boats, if the batteries go dead, you're simply out of luck—you can't walk home!

The electrical system is one area where our automotive counterparts have the advantage. You bring your car into the repair shop, and the technician hooks it up to a diagnostic center. This electronic masterpiece produces all kinds of useful data, everything from which cylinder is misfiring to the amount of poisonous emissions coming out your tailpipe. It will also evaluate the performance of your battery, charging system, and starter motor circuits.

Unfortunately, it's impossible to drive your boat into the diagnostic bay at your local auto shop and get the same sort of checkup. But let's face it—most boatowners today expect this same level of efficiency and technical capability. After all, if they can get it for their $40,000 Lexus, why not their $400,000 Sabre sloop?

CATALYSTS FOR CHANGE

Enter the microprocessor. As we all know, microprocessors are continually becoming smaller and more powerful. With these improvements have come changes in instrument technology that are closing the gap between marine and automotive technicians. Instrument manufacturers today can produce smaller diagnostic equipment with increased functionality, and at prices that many boatowners and marine technicians can afford, or—in the case of the pros—cannot afford to be without.

In addition, advances in battery technology have driven the development and modification of testing instruments. In the old days, we removed the caps from the battery cells and visually inspected the fluid level of each cell. Then—if we had the tools—we used a hydrometer to measure the specific gravity and determine each cell's state of charge, and a carbon pile load tester to determine the battery's ability to supply amperage. Or if we didn't own a carbon pile load tester, we used a voltmeter and checked the voltage across the battery posts while cranking the engine or running the anchor windlass.

These days, however, the vast majority of boats have sealed batteries. These may be either liquid electrolyte batteries or gel-cells and absorbed glass mat (AGM) batteries (referred to as *immobilized electrolyte* batteries by the ABYC). Either way, a hydrometer is of no use because there's no cap to remove. And while a voltmeter will safely determine the battery's state of charge, it won't tell you anything about the battery's ability to deliver amps. Neither will it tell you which cell is bad. But that makes little difference because you can't replace individual cells. One bad cell means you have to replace the battery.

Additionally, load tests on sealed batteries are extremely dangerous. Although they are referred to as sealed batteries, they are more accurately termed *sealed valve regulated* (SVR) batteries. This means that the battery can vent itself if it develops excessive internal pressure—for example, due to an overcharging condition. If this condition goes unchecked, the electrolyte level will eventually fall below the tops of the battery plates, which can cause the battery plates to short-circuit. Conducting a load test on a battery under these circumstances can produce an electrical arc inside the battery case, which now has, as a result of the overcharging, a hydrogen-rich atmosphere. The result? Your battery explodes!

But even if you could use a battery load tester, would you want to? The typical model is heavy and cumbersome to get down into the battery compartments on many boats. It's so much easier to use a handheld device.

Safety, value, functionality, and convenience—all are sound reasons for using the new, alternative instruments and methods that have been developed for battery testing.

CONDUCTANCE BATTERY TESTER

Several years ago, in preparation for an article for *Professional Boatbuilder* magazine, I researched a new breed of conductance battery testers. At the time, I was concerned about the misconceptions and tales of woe I was hearing regarding these instruments, yet I was also intrigued by the possibility that they could be more accurate than traditional battery testing methods. I had only superficial experience with them and had never looked at how they worked. The research was eye opening. Since then I've learned a lot about these devices and have come to believe in their capabilities.

A conductance battery tester is easy and safe to use. No load is placed on the battery, but rather the tester sends low-level, pulsating voltage through the battery from the positive to negative post. By timing this low-level signal, the tester's microprocessor determines the internal resistance, translates that into conductance, and calculates the condition of the battery based on the user-programmed specifications for the battery. This is all accomplished in a matter of seconds, and generally the battery does not

CONDUCTANCE AND IMPEDANCE

Conductance is a measure of the ability of a battery to carry current. It is the inverse of *impedance*, which is a battery's internal resistance. As a battery ages, its internal electrical resistance increases due to sulfate coating on its plates and general deterioration. The result is that impedance increases and conductance decreases.

If you know your battery's original specifications, you can program them into a battery conductance tester to determine the battery's state of health.

CONDUCTANCE BATTERY TESTERS

Micro500XL, Midtronics, www.midtronics.com
DBA Analyzer, Newmar Power, www.newmarpower.com
BAT 121, Bosch, www.boschautoparts.co.uk
CRT-300, Alber, www.alber.com
Accuracy Plus, OTC, www.otctools.com
YA2624, Snap-on, www.snapon.com

need to be recharged before conducting the test (a prerequisite for traditional load testing and specific gravity tests with a hydrometer).

Conductance battery testers, such as those listed in the sidebar, eliminate the need for hydrometers and high-current-draw load testers. They are the only safe means of determining the condition of sealed batteries. When selecting a conductance battery tester, consider the types of batteries it can effectively analyze (e.g., flooded cells, gel-cells, AGM) and the maximum battery rating in cold cranking amps (CCA). Also be sure that the tester matches the battery's capacity. Marine batteries often have a larger capacity than automotive batteries (an 8D-sized battery, for example), which some automotive testers cannot handle.

Midtronics is one of the leading companies involved in the development of these instruments; I use a Midtronics Micro500XL conductance tester, which costs about $800, and have found it to be reliable. It has an amperage rating that accommodates 8D-sized batteries, which are common to the marine

To measure conductance, a tester sends a small signal through the battery, then measures a portion of the pulsed current response. This measures the plate surface available in the battery, which determines how much power the battery can supply.

The Midtronics Micro500XL analyzer with IR-linked printer.

industry. In addition, the unit's long cables—a little more than 10 feet (3 m) in length—have proven quite handy for accessing hard-to-reach batteries. The instrument is versatile, and can also be used for testing starter circuits and charging systems. (All three capabilities will be covered in this chapter.)

The menus on the tester's LCD screen prompt you through the test sequence, and once the testing is done, you can transfer the data via an infrared link to a mini printer. This is handy if you are a professional technician as you can create a report specific to your customer, including your company name and contact information (all user programmable). This printed record is just one more step toward the type of professionalism boaters have come to expect.

Battery Test

You can learn how to use a conductance battery tester in just a few minutes. To perform this test, be sure the battery is connected to the boat's electrical system. If your boat has a multiple battery configuration, test each battery separately (see illustration).

In the photo sequence that follows, the tester is connected to the battery on a boat, and the complete test sequence is shown.

Wiggle the clamps attached to the battery terminals to ensure a tight connection. These meters are extremely sensitive and require a good connection. The Micro500XL will prompt you to do this if it senses a loose or low-quality connection.

The meter has identified a loose connection to the negative battery post clamp. Wiggle the clamp and retest.

low-level pulsating voltage to determine internal resistance

DC +

switch open

fuse

battery

DC –

battery

sub-main breaker

panel-board

Micro500

fused equipment

Leave battery connected to boat's electrical system, but test each battery separately. USE SWITCH TO SELECT BATTERY.

Connect the meter's red clamp to the battery's positive terminal and the black clamp to the negative terminal. You can isolate multiple battery setups by using the battery selector switch installed on the boat. For really large battery banks (with more than two batteries), isolate the batteries by disconnecting the positive lead from each battery you are testing in the system to ensure that the Micro500XL is collecting data for just the one battery and not averaging its readings for several batteries.

The test indicates that the battery's open-circuit voltage is 11.77 V. This is less than 25% of a full charge, as indicated in the graph.

The graph shows the correlation between open-circuit voltage and battery state of charge.

First identify the location of the battery: is the battery in or out of the vehicle (or in this case, the boat)?

Select the type of battery: flooded or AGM. For gel-cells, select "regular" for the best results. (In this case, "flooded" is synonymous with "regular.")

Use the scroll keys to select the battery's rating standard. The typical choices are CCA (cold cranking amps), CA (cranking amps), MCA (marine cranking amps), or DIN (Deutsches Institut für Normung, a German industrial standard). Other options may be JIS (Japanese Industrial Standard) and amp-hours (Ah).

Adjust the numerical amperage value to match the battery rating. (This rating should be on the battery label.)

A typical battery label showing CCA and MCA ratings.

The meter displays the battery's state of health in graph form. A reading of less than 80% indicates you should replace the battery fairly soon; at 60% or below, you should replace it immediately. In this case, the situation looks grim, since the battery is barely giving a reading.

The meter displays the state-of-charge (SOC) test, the results of the CCA test, and the battery's CCA rating. In this example, the 600 CCA battery tested at 149 CCA, with a state of charge of 11.78 V. Based on these results, the unit's recommendation is to "charge & retest" the battery (top of screen). (Note: In some test situations, the battery being tested may have such a low state of charge that the results will be skewed, confusing the battery tester's microprocessor. In this case, the unit's recommendation will be "charge & retest.")

Starter Circuit Test

The Micro500XL can perform a complete starter circuit test, including measuring for minimum cranking voltage. Attach the red and black leads to the cranking battery as you did for the battery test (red to positive and black to negative), then follow the meter's instructions (see photos page 20).

Note that the cranking voltage test is only one of several tests that can be used to troubleshoot starter circuit problems. Current draw tests and voltage drop tests are also frequently used. We'll look at how these are done using another high-tech tester later in this chapter.

At the end of the test, the unit can generate and print a report. Make sure that you disconnect the battery before printing.

The initial screen for the starter circuit test: first, start the engine. (Note: If you're testing an older diesel engine in cold temperatures, warm up the engine for 5 minutes before beginning this test to prevent inaccurate results.)

This next screen shows the results. In this case, the starter circuit in question brought the system voltage down to 10.89 V, a normal and acceptable condition. (Anything less than 10.5 V on a computer-controlled engine is considered unacceptable. On older engines without computer controls, a minimum of 9.6 V is acceptable.)

MINIMUM CRANKING VOLTAGE AND VOLTAGE SPIKES

Minimum cranking voltage is especially important on today's modern boats. The computer-driven systems found aboard virtually all boats, as well as common marine electronic equipment such as fishfinders and GPS chartplotters, are all sensitive to the low-voltage conditions that can occur during engine cranking. In fact, low voltage during cranking affects any system that shares the same battery(s) as the starter.

What is the minimum? Most modern engine manufacturers recommend nothing less than 10.5 VDC. In most cases, voltage levels below 10.5 VDC will increase the possibility of an onboard computer "blinking out" or losing data, as well as cause a variety of engine symptoms, such as excessively rich fuel mixtures, rough running, or a complete failure to start. Symptoms will manifest differently from one engine or system to another, but suffice it to say that cranking voltage is of more concern now than ever.

High voltage spikes are also a concern, and are sometimes caused by solenoids for battery combiner systems and the like due to *inductance* (a property of a conductor or coil that determines how much voltage will be induced in it by a change in current) as these devices open and close circuits. This problem is dealt with most effectively by electrically isolating electronic navigation circuitry from cranking motor circuits, which is often done by using a completely separate battery supply for the electronic equipment.

Charging System Test

The Micro500XL can perform several important charging system tests, as shown in the following photo sequence. As you will see, the unit instructs you at each step. (As with the starter circuit test, if you are testing an older diesel engine in cold temperatures, first warm up the engine for 5 minutes.)

Trend Analysis

One way to use the Micro500XL is to predict when a battery is going to give up the ghost. You may remember that in photo 11 on page 19, the screen display showed the rated CCA and the measured CCA in real time based on the

Press the ENTER key on the face of the instrument to begin the charging system test.

Turn off all electrical loads and rev the engine for 5 seconds. The unit provides a screen view to ensure that engine rpm is detected. It then advances to a horizontal tachometer-type display. Bring the engine rpm to the level of the vertical line and hold.

Now turn on electrical loads. Select critical, commonly used DC loads that you may need or use simultaneously with the engine running; for example, navigation lights, bilge pump and blower circuits, all navigation and communications equipment, and refrigeration systems. You want to be certain that the system can supply all these circuits and have adequate power left to recharge the batteries at the same time.

The rpm scale for the loaded running test. Rev the engine to the vertical line and hold.

With loads on, idle the engine.

After idling, rev the engine again to the vertical line. This screen shot shows the engine is just below the correct rpm.

The diode/ripple test measures AC voltage leakage past the rectifier assembly in the alternator. This test can indicate imminent charging system failure, or an electrically "noisy" alternator that can affect the performance of some onboard marine electronic equipment. This test is simply a go, no-go test—pass/fail. It can be performed separately with a digital multimeter set to the AC volts scale (anything in excess of 0.4 VAC is considered excessive).

At the end of the charging system test, the Micro500XL provides an overall analysis of the charging system. In this example, the system is normal.

This screen displays the actual voltage for the loaded and unloaded tests. As you can see, the voltage reading in the unloaded test is higher (by 0.13 V) than the voltage at the battery terminals with the engine turned off. This difference just means the charging system is doing something. However, if the unloaded voltage is more than 2.5 V above the initial battery reading with the engine off, then the voltage regulator is at fault. If the loaded voltage is less than 0.5 volt above the static battery voltage, the alternator's amperage capacity is too small for the loads it is serving.

results of the battery condition tests. This information has more value than you might think, but it requires a bit of prior recordkeeping on your part.

When you install a new battery, perform a conductance test on the first day of the battery's service life and log the CCA. This measurement will be your starting point, or benchmark. Then periodically test the battery and log the results. Generally, batteries don't fail suddenly, but deteriorate gradually. As your battery ages,

your log entries will show a gradual decrease in available CCA. By tracking the CCA lost over time, you will establish a trend based on your boat's actual use. This will enable you to make an educated guess as to when battery failure is imminent. The magic number is easy to remember—look for 80% of the battery's rated CCA value. (This is not an arbitrary value; it is based on standards established by the Institute of Electrical and Electronics Engineers, IEEE.) At that point, it's time to replace the battery if you want to ensure its continued reliability.

DIAGNOSTIC METER

Although clearly an automotive-based instrument, a diagnostic meter is easily adapted to marine uses and provides added functionality and increased accuracy over a battery conductance tester. The Midtronics inTELLECT EXP-1000, which costs about $1,100, has functions for battery, starter circuit, and alternator testing, as well as capabilities for voltage drop testing. (As of this writing—2007—I know of no other units with voltage drop testing capability via conductance testing.) With the inTELLECT EXP-1000, you can isolate circuit faults caused by loose, corroded connections or undersized wiring—two common problems on boats—and it can also function as a multimeter.

Features of the inTELLECT EXP include:

- Comprehensive battery tests, with an improved algorithm (compared to the Micro500XL) for more accurate tests on deep-cycle batteries as well as AGMs and gel-cells.

- Capability to create performance graphs for such things as starter current draw and alternator output. These features can be extremely useful in analyzing marginal performance situations.

- A simple cable voltage drop test, performed by attaching the leads to each end of any circuit. It will definitively isolate excessive voltage drop to either the power feed or ground return side of the circuit. This saves having to check each side of the circuit separately using a conventional multimeter, and will provide a pass/fail answer based on user-programmable parameters entered into the meter before testing begins.

- Full digital multimeter capabilities, including DC volts, AC volts, amperage (with the optional amp clamp), temperature, ohms, and test diodes. It also provides a small oscilloscope.

- Battery-operated printer connected to the unit via an infrared link. This lets you print full test results as a record for customers or as backup to a work order for future reference.

As described in my book *Powerboater's Guide to Electrical Systems*, voltage drop tests previously required that the circuit in question be activated and that at least some current be allowed to flow through it. This can be somewhat of a handicap on circuits that have simply stopped working. The inTELLECT EXP-1000 provides its own low-level power to the circuit to enable its conductance testing capabilities, making it possible to determine if excessive voltage drop exists without having a live circuit. Further, the inTELLECT EXP-1000 will automatically determine the exact voltage drop in volts, and isolate whether the excessive drop is on the power (supply) side or the negative (return) side of the circuit. This saves considerable time over the traditional methods used to determine the location of excessive voltage drop in a circuit.

Battery Test

The inTELLECT EXP represents the latest evolution in the Midtronics line of battery testers. Like the Micro500XL, it uses conductance

The inTELLECT EXP-1000 has a large screen display, an alphanumeric keypad, and a number of unique capabilities.

The main menu of the inTELLECT EXP-1000.

FULL-FEATURE DIAGNOSTIC METER

inTELLECT EXP-1000, Midtronics, www.midtronics.com

First, you enter some basic information, such as whether the battery is in or out of the "vehicle."

Then you enter the type of battery connection: top post, side post, or "remote." This last option accommodates testing from the "jump start" terminals now incorporated on some over-the-road vehicles, and even some new boats. You can use this option to hook into the positive and negative DC bus bars found behind most boat electrical panels, eliminating the need to crawl into some tight battery installation. Batteries connected in series or in parallel should be tested in isolation from one another, however, so unless it's possible to isolate the batteries by switching, the remote connection option may not be practical. The unit's red lead goes to the battery positive terminal; the black lead to the negative.

Next, you enter the battery type: regular flooded, AGM, AGM spiral wound (such as the Optima brand), or gel-cell.

Then you enter the battery capacity rating standard shown on the battery label. Eight choices are provided (four are shown), covering all world standards for battery ratings.

Then you enter a specific rating value. In the case of CCA or MCA, the entry range is from 100 to 3,000 amps. This represents more than enough capacity to deal with even the largest marine battery in common use, the 8D, which has a CCA rating between about 1,200 and 1,600 amps, depending on the make and model. After entering the required data, press "Next" and the test begins.

The results of the battery condition test: Good Battery. At 12.48 V, the battery is not fully charged, but because its indicated capacity of 543 CCA is over its rated capacity, the battery is still in good condition.

accurately diagnose problems with all three common battery types: flooded (liquid electrolyte), gel-cells, and AGMs. The photo sequence shows the battery test procedure.

If conditions are outside of the parameters—such as low temperature or deeply discharged—required for an accurate test, one of several screen tests may appear next:

- Temperature compensation test: The inTELLECT EXP-1000 may ask you to measure the battery's temperature with the unit's infrared temperature sensor if the unit detects that the temperature may make a difference in the result. It will prompt you to place the sensor within 2 inches (50 mm) of either the top or side of the battery. After the temperature reading is displayed, hit the "Next" key. The instrument will capture the temperature data and recalculate its findings.

- Deep scan test: If temperature compensation doesn't allow the in TELLECT EXP-1000 to complete its calculations, the unit will jump to a deep scan test to further analyze the battery. It may ask you to run a

The SOC screen presents battery voltage from photo 6 in an easily comprehended graphic format. In this case, it tells you that 12.48 V is 75% SOC.

technology to determine battery condition, only with a more sophisticated algorithm programmed to its internal microprocessor. This subtle improvement allows the tester to more

TABLE
2-1

Battery Decisions and Recommendations

Decision	Recommended Action
GOOD BATTERY	Return the battery to service.
GOOD–RECHARGE	Fully charge the battery and return it to service.
CHARGE & RETEST*	Fully charge the battery and retest. **Failure to fully charge the battery before retesting may cause false readings.** If CHARGE & RETEST appears again after you fully charge the battery, replace the battery.
REPLACE BATTERY	Replace the battery and retest. A REPLACE BATTERY result may also mean a poor connection between the battery cables and the battery. After disconnecting the battery cables, retest the battery using the out-of-vehicle test before replacing it.
BAD CELL–REPLACE**	Replace the battery and retest.

*If the result is CHARGE & RETEST, the EXP will calculate and display the time needed to charge the battery at 10, 20, and 40 A.
** When testing at the remote posts, the EXP may need to verify the result. It will give you the option of retesting at the battery posts.

Source: Midtronics

Finally, the meter displays a state of health (SOH) screen, a summary based on the measured CCA versus the programmed or rated CCA. In this case, the meter "dial" shows the pointer between 50% and 100%, indicating approximately 70% to 75% state of health. This battery will need to be replaced fairly soon. The magic number to remember is 80% of the rated CCA. Virtually every battery I've ever tested using a conductance tester will measure more than its rating when fairly new. Once a battery measures less than 80% of its rating, it should be replaced. It's important not to confuse state of charge with state of health: SOC is merely a reading of the open-circuit voltage across the battery's terminals, but really doesn't provide any information about the battery's ability to deliver current as SOH does.

5-minute discharged battery test, either before or after recharging the battery.

At the end of the testing, the inTELLECT EXP-1000 displays one of five test result messages: good battery; good but recharge; charge and retest; replace battery; bad cell—replace. Depending upon the results, numerical values are displayed, as are two screens showing battery state of charge (SOC) and state of health. These last two screens are user selectable.

Starter Circuit Test

If you initially chose "system test" from the main menu, the inTELLECT EXP-1000 will move automatically to the next test—the starter circuit test.

To get the most comprehensive results, you will need the optional amp clamp. Place the clamp around the main starter battery feed cable to measure the amperage the starter is drawing while it is cranking. This is an important part of overall starting system analysis. Assuming that the engine isn't seized, an excessively high amperage draw indicates a possible fault inside the starter motor. An excessively low amperage draw indicates a faulty connection to the starter, which could be caused by either a loose or corroded connection or cabling that is too small for the job.

While the inTELLECT EXP-1000's optional clamp is recommended, be aware

Results of the cranking motor test, showing voltage, amps, and cranking time. (With an attached amp clamp, this screen would also display the starter loop resistance reading.)

TABLE 2-2	Starter System Decisions and Recommendations	
Decision	**Action**	
CRANKING NORMAL	The starter voltage is normal and the battery is fully charged.	
LOW VOLTAGE	The starter voltage is low and the battery is fully charged.	
CHARGE BATTERY	The starter voltage is low and the battery is discharged. Fully charge the battery and repeat the starter system test.	
REPLACE BATTERY	(If the battery test result was REPLACE or BAD CELL.) The battery must be replaced before testing the starter.	
LOW CRANKING AMPS	The starter voltage is high but the cranking amps are low.	
NO START	The engine didn't start and the test was aborted.	
CRANKING SKIPPED	The EXP didn't detect the vehicle's starting profile and skipped the Starter Test.	

Source: Midtronics

that its measurement range (up to 700 amps) is not enough to capture starter current draw on some modern medium and large marine diesel engines. If you frequently work with such engines, you might want to invest in a stand-alone amp clamp with a range up to 1,000 amps DC. (If you are not using the inTELLECT EXP-1000's clamp, select "not available" during the initial test setup.)

First, the inTELLECT EXP-1000 verifies the battery condition (described above in the Battery Test section) as a prerequisite to any starter circuit test. Once a good battery has been confirmed, completing the test procedure is simply a matter of starting the engine. Use the same, simple two-wire positive and negative connections at the battery already in place and connect the optional amp clamp around the battery negative cable.

With these in place, the inTELLECT EXP-1000 will measure and record the following:

- average cranking voltage
- cranking amperage (if you use the optional amp clamp)
- cranking time in milliseconds
- starter loop resistance (in ohms); note that this test will only be useful if you

The voltage reading during the cranking cycle.

have previously established a baseline for the circuit, as with cranking amps

Normal figures for older, noncomputerized engines must provide a reading of over 9.6 V while cranking. Modern engines should always read in excess of 10.5 V.

Interpreting the Results

Voltage and amperage readings can tell you many things beyond whether the starter circuit is in order. Particularly on modern boats, which are increasingly powered by electronically controlled engines, excessive system voltage drop during engine cranking is a concern. For example, assuming the connected battery has tested OK, voltage readings that are lower than desirable may indicate that the battery is too small (i.e., not enough capacity) for the task at hand, or that the wiring is too small for the job and an upgrade is required.

Cranking amperage is also of concern, as mentioned above, but historically, engine manufacturers have been more than a little shy about sharing relevant cranking amp specifications with technicians or end users of their products. In the old days, 1 amp per cubic inch of engine displacement was the rough rule of thumb, but this applied only to gasoline engines. Diesel engines, with their much higher compression ratios, often pulled far more than 1 amp per cubic inch.

Starter motor technology has changed, too. The 1 amp figure was reasonable for old-style field-wound motors, which characteristically pull more amperage. But with the evolution to permanent magnet–type motors, gear reduction motors, and such, starter current draws have decreased dramatically in the last ten to fifteen years.

Therefore, as with other measurements, starter current draw as a diagnostic tool is only as good as the information you start with. Do you know what the draw is supposed to be? If you are working on the same systems repeatedly,

or the starter in question is on your own boat, establish a benchmark when everything is functioning normally. Log this number and the temperature at which the value was measured for future reference. Remember, cold temperatures will increase the load on the starter and cause the motor to pull more amperage. Typical draws for modern starter circuits range anywhere from 150 to 850 amps at 80°F (27°C). At 80°F, any variation from the norm that exceeds about ±10% to 20% indicates a problem.

Higher than normal current draw could indicate a seized engine, although this should have been ruled out already. (Remember, you are probably performing these tests to establish the cause for slow engine cranking. A partially seized engine is generally the result of an extreme overheating incident, in which case the temperature gauge would have registered over 215°F (102°C). The vessel's history and other symptoms will usually confirm or deny this possibility.) Assuming the engine is in good condition, higher than normal current draw is probably a problem internal to the starter motor. Remove the starter for further inspection and repair as needed.

Lower than normal current draw is a classic symptom of loose or corroded cable connections or undersized wiring (which will show on the inTELLECT EXP-1000 as higher than normal loop ohms resistance). You may need to conduct additional testing to determine the location of the excessive resistance within the circuit. If the wiring checks out OK, then the problem is within the motor itself.

Charging System Test

The third test under the System Test (main menu) is the charging system test. As with the other tests, it is easy to perform by following the unit's on-screen prompts. The cable connections are the same as for the battery and starter tests, with the possible addition of the amp clamp. (If you are following the meter's system test procedure, advance to the charging

The sequence begins with the unit prompting you to turn off all loads and idle the engine. If the unit senses a no-charge condition at idle (common to many diesels with self-exciting alternators), it may ask you to confirm whether a diesel is the subject of the test. (Some diesel engines require the engine to be revved at above idle speed to excite the alternator into producing a charge.)

Next the unit prompts you to rev the engine for 5 seconds to a certain rpm level (indicated by a hash mark) with no electrical loads. The EXP-1000 will tell you when engine rpm is detected; at that point, hit the advance key on the meter to move to the data page, which will give the results of the test with loads off. Then it tests for problems, such as excessive AC ripple, voltage output, and charging amperage values if the amp clamp is used.

system test by pressing the "Next" key after the starter circuit test is completed.)

AC ripple is the amount of AC leakage past the diodes in the alternator's rectifier assembly, which indicates either alternator stator damage or, more likely, rectifier diode failure or very low quality diodes. I generally use 0.4 VAC as the maximum acceptable reading. More than that will almost surely induce excessive RFI (radio frequency interference), with the potential for interfering with electronic systems on board. (Many of the worst cases of excessive AC ripple I've tested in the field were on newly rebuilt alternators, which says something about the quality of the diodes some remanufacturers use.)

For the next phase of the test, turn on electrical loads. The meter suggests automotive accessories such as headlights and the heater blower fan, but you should switch on the critical loads as defined in ABYC Standard E-11 for DC load calculations (see column A in the ABYC worksheet on page 30). Some typical DC amperage draw values for common onboard loads are shown in Table 2-3.

TABLE 2-3	DC Current Draw
Equipment	**Load (amps)**
Anchor light	1–3
Bilge blower	1–8
Bilge pump	2–8
DC refrigeration	5–20
Depth sounder	0.1–4
Knotmeter	0.1–1
GPS/chartplotter	0.1–5
Masthead light	1–2
Radar	4–8
Running lights	3–6
SSB radio	1–30 (30 in transmit mode)
VHF radio	0.7–1.5
Windshield wiper motors	1.5–5 (each)

The values given in Table 2-3 are approximate, and actual current draw will vary based on specific equipment. It's always best to use an inductive amp clamp and create your own log of current draw for every circuit on a boat. This log can become a useful diagnostic aid when problems do arise.

Measuring current draw with an inductive amp clamp is not too time consuming. Before testing any appliance for current draw, disconnect the appliance from the boat circuitry and connect it directly to a good battery with a pair of jumper cables. (The gauge of battery jumper cables is usually sufficient to carry all the needed current and obtain valid readings.) Set the inductive amp clamp around the DC positive conductor to measure the current draw and log the value. This will provide the benchmark value to effectively use the inTELLECT once the appliance is reconnected to the boat's circuitry; if the inTELLECT indicates anything less than the benchmark value, there is excessive voltage drop in the circuitry. This could be due to a loose or corroded connection or a wire that is too small for the job.

A	AMPERES	B	AMPERES
Navigation Lights		Cigarette Lighter	
Bilge Blower(s)		Cabin Lighting	
Bilge Pump(s)		Horn	
Wiper(s)		Additional Electronic Equipment	
Largest Radio (Transmit Mode)		Trim Tabs	
Depth Sounder		Power Trim	
Radar		Toilets	
Searchlight		Anchor Windlass	
Instrument(s)		Winches	
Alarm System (standby mode)		Fresh Water Pump(s)	
Refrigerator			
Engine Electronics			
Total Column A		Total Column B	
		10% Column B	
		Largest Item in Column B	

Total Load Required
Total Column A   _____
Total Column B   _____   (The larger of 10% of Colum B or the largest item)
Total Load   _____

The electrical load requirement worksheet from ABYC Standard E-11. (Courtesy ABYC)

Turn on critical electrical loads as specified by ABYC Standard E-11, as shown in the worksheet opposite. After a few seconds the EXP will prompt you to rev the engine for 5 seconds with the loads on. The correct rpm level will be shown in a graphic display similar to that shown for the no-load test in photo 2 on page 29.

Next, turn off the loads and the engine.

The arrow keys allow you to scroll through a series of results in bar graph format.

The test results. In this case the system only lost 0.01 V between the loaded and unloaded test, indicating that the alternator charging system had no problem keeping up with the loads. The minimum to look for is 0.5 V more than the initial open-circuit voltage reading on the battery before the alternator is even operating (i.e., with the engine not running). The EXP-1000 will check this automatically; thus the screen displays "no problems."

The results of the AC ripple test, showing peak-to-peak AC voltage and a graph of the diode waveform.

TABLE 2-4	Alternator Decisions and Recommendations

Decision	Action
NO PROBLEMS	The system is showing normal output from the alternator. No problem detected.
NO OUTPUT	The alternator is not providing charging current to the battery. • Check the belts to ensure the alternator is rotating with the engine running. Replace broken or slipping belts and retest. • Check all connections to and from the alternator, especially the connection to the battery. If the connection is loose or heavily corroded, clean or replace the cable and retest. • If the belts and connections are in good working condition, replace the alternator. (Older vehicles [and modern boats with high-output alternators often] use external voltage regulators, which may require only replacement of the voltage regulator.)
LOW OUTPUT	The alternator is not providing enough current to power the system's electrical loads and charge the battery. • Check the belts to ensure the alternator is rotating with the engine running. Replace broken or slipping belts and retest. • Check the connections from the alternator to the battery. If the connection is loose or heavily corroded, clean or replace the cable and retest.
HIGH OUTPUT	The voltage output from the alternator to the battery exceeds the normal limits of a functioning regulator. • Check to ensure there are no loose connections and that the ground connection is normal. If there are no connection problems, replace the regulator. (Most alternators have a built-in regulator requiring you to replace the alternator. In older vehicles [and some modern boats with high-output alternators] that use external regulators, you may need to replace only the voltage regulator. The regulator controls voltage ouput based on the battery voltage, underhood [or engine room] temperature, and vehicle loads used. In other words, it controls the maximum voltage the system can produce based on the current needs and amount of current that can be produced by the spinning of the rotor in the alternator. The normal high limit of a typical automotive regulator is 14.5 volts ±0.5. Refer to the manufacturer specifications for the current limit, which may vary by vehicle type. A high charging rate will overcharge the battery and may decrease its life and cause it to fail. If the battery test decision is REPLACE and the charging system test shows a HIGH OUTPUT, check the battery's electrolyte levels. A symptom of overcharging is battery fluid spewing through the vent caps, which causes low electrolyte levels and will harm the battery.
EXCESSIVE RIPPLE	One or more diodes in the alternator aren't functioning or there's stator damage, which is shown by an excessive amount of AC ripple current supplied to the battery. • Make sure the alternator mounting is sturdy and that the belts are in good shape, and functioning properly. If the mounting and belts are good, replace the alternator.
OPEN PHASE	The EXP has detected an open phase within the alternator. Replace the alternator.
OPEN DIODE	The EXP has detected an open diode within the alternator. Replace the alternator.
SHORTED DIODE	The EXP has detected a shorted diode within the alternator. Replace the alternator.

Source: Midtronics

If you use the optional amp clamp, the unit will display alternator output in amps, both loaded and unloaded. This information is useful for performing a load analysis. The alternator's job is to supply voltage to normally running electrical equipment *and* have at least 0.5 V left over to recharge the battery. An underrated (in amps) alternator will not be able to do this and will display less than the 0.5 V increase over static voltage with loads turned on. Again, you can perform this test with a stand-alone amp clamp around either the battery positive or negative cable.

Voltage Drop Test

I've been endorsing voltage drop as a way to pinpoint weak links in an electrical circuit for years, but there is still a general lack of understanding of the procedure and its applications. Many troubleshooters set their digital volt-ohmmeter (DVOM) to the ohms scale and check for continuity at various points throughout a circuit. Others use a 12 V test light to see if there is enough potential at a given point in a circuit to light the bulb. Neither of these approaches is really adequate, however. An ohmmeter may show continuity, and even a measurable resistance value, but what does that tell you if you don't know what the resistance *should* be? The bulb in the test light may glow, but is it glowing brightly enough? Only by measuring voltage drop can you really get a handle on the quality of a circuit and all its termination points.

To measure voltage drop—i.e., how much voltage is "lost" in a circuit—there has to be at least some current flow. At one time, this test couldn't be performed on an inactive circuit, but as noted earlier, the EXP generates its own signal, so this is no longer a problem.

Significance of Voltage Drop

Keeping in mind that conductance is the inverse of impedance (resistance), excessive voltage drop is caused by high electrical resistance. Too much

TABLE 2-5	Acceptable DC Voltage Loss (per ABYC E-11)		
System Voltage	Critical Circuits (3%)		Noncritical Circuits (10%)
12	0.36 V		1.2 V
24	0.72 V		2.4 V

resistance (which is measurable) can cause a circuit to perform poorly or not at all. ABYC Standard E-11 identifies key circuits and the acceptable levels of voltage drop as a percentage of nominal voltage. For critical circuits like electronics, bilge blowers, panel feeders, and navigation lights, the limit is 3% of the nominal circuit voltage (generally either 12 or 24 V). For noncritical circuits, such as cabin fans and electric heads, the figure is 10%. These translate into the specific voltage losses shown in Table 2-5.

These numbers represent the maximum amount of voltage we want to see "go missing" due to resistance in electrical circuitry, exclusive of the actual load in the circuit. (The lost voltage actually turns up as heat, the primary by-product of excessive electrical resistance.) These values, I believe, are part of the reason why many electricians have trouble grasping the significance of voltage drop: the numbers seem too small to worry about. But in fact, these small numbers have great significance. Losses in excess of these amounts are due to undersized wiring or loose or corroded termination points—among the most common problems in DC marine systems. (We usually don't worry too much about AC voltage drop on board because the wire runs in AC circuits are generally too short to be a major factor.)

Using the inTELLECT EXP-1000

The inTELLECT EXP-1000 generates a low-level signal and calculates the voltage drop over a given leg of the circuit or over the entire circuit, depending upon where you place the leads. The readout identifies the voltage drop in the positive and negative sides of the circuit as separate values, which should save you a lot of time in narrowing down the exact location of the fault. (It's usually at one end of the circuit or the other, or any terminal point in between.)

To calculate voltage drop, both of the inTELLECT EXP-1000's lead sets are required. The battery test lead and the DMM (digital multimeter) lead have different plug ends, so you can't connect them incorrectly at the meter. Each lead has one black and one red end clamp for attachment to the circuit.

Attach the red lead from the DMM lead on the positive side at the circuit's source of power, and the black lead to the negative side. At the other end of the circuit, connect the red battery lead to the positive connection, and the black lead to the negative connection.

After entering the amperage draw value for the circuit, hit the "Next" key. The device will test the circuit and display total circuit voltage drop. It will also display individual values for both the positive and negative return sides of the circuit, saving several steps compared to the use of a traditional multimeter.

Using a conventional meter, you have to connect, disconnect, and reconnect both ends of the circuit to acquire the same data. Then you would still have to perform some calculations to get voltage drop.

The results screen for a tested circuit. It shows the total voltage drop and the separate drops on the positive (0.04 V) and negative (0.11 V) sides of the circuit. These are typical low-voltage readings and well below the 3% maximum value you should see at the main power supply circuit.

Limitations of the inTELLECT EXP-1000

Despite the inTELLECT EXP-1000's tremendous capabilities, it does have some minor limitations:

1 You are limited by cable length. The length of circuit you can test is only about 18 feet (5.5 m), which is the combined length of the unit's two cables. Midtronics states that extending the cables can result in a loss of signal strength, which might affect the microprocessor algorithms. Happily, 18 feet is sufficient for most circuits.

2 You must know the circuit's amperage. The EXP requires you to enter the circuit's amperage value. Acquiring the amperage value for a given circuit means you have to know the current rating for a particular appliance in the circuit. Look for the rating on a plate on the appliance

or in the manufacturer's installation manual or specification sheet.

Note: It's a good idea to log this information and update it when you add or swap a piece of AC or DC equipment. You'll be prepared for future circuit troubleshooting.

3 If the correct current draw is not known, some preliminary work is needed to make sure the inTELLECT is programmed for the correct amperage (see page 30).

VALUE AND UTILITY

In addition to all the above-mentioned functions, the inTELLECT EXP-1000 can also act as a simple multimeter for general electrical work around the boat—a handy feature, although hardly enough on its own to justify the investment in this type of instrument.

In my opinion, the inTELLECT EXP-1000 provides more functionality than the Micro500XL. Although the Micro500XL performs a good battery test, it may not be as precise in its evaluation of very large (8D-sized) AGM and gel-cell batteries. Additionally, the inTELLECT EXP-1000's voltage drop testing capabilities are a major breakthrough.

The typical boatowner would be served well by the Micro500XL, but for the advanced professional technician, I strongly recommend the inTELLECT EXP-1000. It can generate useful troubleshooting data more readily than any other method. It may, however, give the average boatowner more information than he or she needs.

Although I am not aware of other devices similar to the inTELLECT EXP-1000, conductance battery testers comparable to the Micro500XL are available from a number of sources, including Snap-on and OTC (see page 16). The offerings of these and other companies may be of equal utility and value, but I believe you will benefit from seeing the capabilities of the specific models shown here as these types of instruments are not yet widely known.

The equipment discussed in this chapter is not cheap; indeed, these items may be among the most expensive test equipment in your shop. But when you consider that twenty-five years ago a professional technician might have paid a similar amount for a high-end carbon pile load tester that only tested batteries, you can easily justify the cost in terms of functionality, safety, and payoff potential.

Keep in mind that this field is developing rapidly, and we should expect numerous improvements, new functionalities, and new models and suppliers on a regular basis. When you're ready to buy, take a thorough look at what the market offers and, if you want to remain at the cutting edge as a top-notch technician or thoroughly capable boatowner, buy the best you can afford.

Testing Continuity and Tracing Circuits

Two of the most time-consuming tasks for marine troubleshooters are finding breaks in long circuits and tracing wires through tangled bundles and harnesses that snake in and out of view as they pass through bulkheads and behind cabinetry. Tone-generating circuit tracers and time domain reflectometers (TDRs) are two modern tools that greatly simplify these tasks.

Like many of the tools in this book, tone-generating circuit tracers and TDRs have been used for years in other industries, such as IT and telecommunications, before being "discovered" relatively recently by high-end marine electrical technicians. Few marine technicians, and far fewer private boatowners, have heard of or used these very useful tools.

Without these tools, however, troubleshooters must struggle with a multimeter or DVOM set to the ohms scale, and run continuity checks one wire at a time. This often involves dragging an extended connection lead through the boat to attach both ends of the meter leads to the wire or harness. And even if the continuity test does confirm a break in the circuit, you are still left wondering exactly where the break is located.

TDRs and circuit tracers eliminate all that work. When connected to one end of a pair of conductors (and it must be a pair), both devices bounce a signal between the two conductors. When the "bouncing" stops, a fault is indicated. But the similarities end there.

Working from one end of the conductor, a TDR can also identify exactly how far down the wire run a problem is located. (Amazingly, TDRs have an effective range of 6,000 feet/1,829 m—well beyond the distance needs on boats.) Tone-generating circuit tracers, in contrast, require access to the entire length of the wire. You connect the transmitter to one end of the conductors, then trace along the wire with the handheld receiver unit. As long as the receiver sounds off, it is receiving a signal, indicating continuity. When the noise stops, you've found your break.

If access is not a problem, a tone-generating circuit tracer may be easier to use, since it *shows* you the exact location of the fault. If access is a problem, a TDR may more useful since it *tells* you where the fault is located, as a function of distance from the device. You then have to find the location, perhaps using a tape measure.

Pricing is another major difference. Unlike many of the other tools discussed in this book, tone-generating circuit tracers are an extremely modest investment, running in the

The Wire Tracker tone-generating circuit tracer (left) is sold in the United States by Ancor Marine. It is really two devices: a signal-generating transmitter and a handheld tracer or receiver. The Fault Mapper model CA7024 alphanumeric TDR (right) is made by AEMC Instruments.

Next we'll look at how these tools are used for tracing wires through boats, finding open circuits or breaks in wires, and in the case of the TDR, how far down a wire run the break or short circuit is located.

TONE-GENERATING CIRCUIT TRACER

A tone-generating circuit tracer has two components: a transmitter, and a handheld receiver or tracer. The transmitter, connected by its dual leads to a pair of cables routed together, sends an oscillating signal down the length of one of the conductors. Most units operate in a frequency range between about 800 Hertz (Hz) and 1,100 Hz. This signal bounces, or reflects, off the parallel cable, something like radar. The receiver unit's pointed end detects this reflected signal and amplifies it so that you can hear it. (Most units also have a small light built into the receiver that glows as it senses the signal.) By moving the receiver down the length of a wire run, you can effectively trace the signal across the conductor's entire length. If the receiver suddenly stops producing audible and visual signals, an open circuit or break is implied. Right there you may have found the source of your electrical problem.

$60 to $100 range. TDRs, on the other hand, are not quite as basic. The alphanumeric unit described in this chapter costs about $500. The time it can save in troubleshooting problems, however, more than justifies its price.

Warning: You should trace only unenergized wiring. Contact with live circuits can result in severe injury or death. Always disconnect power to the circuit before using a circuit tracer.

You must have access to the entire wire run and be able to track down the entire length of the harness to find a break. Unfortunately, you

TONE-GENERATING CIRCUIT TRACERS

Wire Tracker, Ancor Marine, www.ancorproducts.com
Fox and Hound, Triplett, www.triplett.com
TR02 Cable Tester, AEMC, www.aemc.com

For this example, connect the transmitter at the source of a circuit on a small fuse panel. Clip the transmitter's red lead to the output, or DC positive, side of the fuse, and clip the black lead to the terminal on the panel's DC negative bus bar. Notice that both of these leads are bundled together in a harness with four other wires as they head away from the panel, so the signal generated from the tool will be able to bounce off the leads as the signal travels the length of the harness.

Follow the harness to another point in the boat. You'll know the circuit is still intact because the tracer/receiver is sending out a loud tone, and the light at the top of the tool continues to glow. If the light goes out and the sound stops, an open circuit is indicated. If you have or can gain access to the harness throughout its run, you can pinpoint the break by tracing back to the point where the tone and light come back on.

will often not have that luxury on many installations because many cable runs are routed through conduits or behind installed cabinetry. That said, the device is useful for identifying which wires are which as they reappear at the opposite end of a hidden run, as long as there are no breaks along the way.

The photo sequence illustrates the Ancor tracker tool in use.

Note: The power must be turned off either at the DC main on the panel or at the battery master switch on the boat when using this tool; it can't share power transmission and its own signal.

Frequently a harness will disappear behind a panel or into a conduit. At the opposite end, use the tone-generating circuit tracer to identify the conductor/pair you've been tracing. As long as the circuit has continuity up to that point, the device will glow and sound off when you touch the receiver unit against the correct wire.

TIME DOMAIN REFLECTOMETERS

Fault Mapper CA7024, AEMC, www.aemc.com
Model 3271, Triplett, www.triplett.com

ALPHANUMERIC TIME DOMAIN REFLECTOMETER

The TDR has been around for some time. Widely used by the military, computer network technicians, and the telecommunications industry, the technology is proven and accepted as the most efficient way to trace problems in long wire runs. Historically, these devices have

Time Domain Reflectometry

origin   fault

transmitted pulse

t_1

reflected pulse

t_2

$$\text{Distance} = \left(\frac{t_1 + t_2}{2}\right)\left(C_o\right)\left(V_p\right)$$

C_o = speed of light in a vacuum
V_p = velocity of propagation

A TDR transmits pulses of a known shape and amplitude into one end of a cable. The pulses travel along the cable at a speed determined by its VOP. As they reach impedance changes in the insulation of the cable, indicating a fault or a cable end, the reflections caused travel back along the cable and are identified by the TDR. The size, shape, and general nature of the reflected pulses indicate the type of fault encountered. The distance to the fault can be determined from the time it takes for the pulse to be reflected.

been expensive; at prices ranging from $5,000 to $10,000, they were beyond the means of almost all marine electricians, and out of the question for boatowners. A few years ago, however, costs dropped dramatically, and good alphanumeric TDRs became available for around $500. I immediately tested and subsequently purchased one of these new-generation units, and I'll never go back to using an ohmmeter for tracing wire runs again. This tool is the only way to go in today's world.

Velocity of Propagation

A TDR needs to know the *velocity of propagation* (VOP) of the wire being tested before it can begin.

VOP is measured as a percentage of the speed of light (186,000 miles per second or 299,792,458 meters per second). A VOP value of 100% simply means that a cable could conduct electricity at the speed of light. But because all cable has some inherent resistance, the VOP value will always be less than 100%, meaning that the electricity is running through the cable at a percentage of the speed of light. And as that percentage will vary with size and type, the TDR must be calibrated for the type of wire or cable you plan to test.

Although the AEMC Fault Mapper, the tool I use, contains a fairly comprehensive library of VOP values for commonly used network and coaxial cabling, none of these come close to the values for commonly used marine wiring. Fortunately, this shortcoming is easily overcome, as described below.

Determining the VOP Value

For this example, we are using the AEMC Fault Mapper's method to determine VOP, but the steps will be similar for other brands:

1 Cut a length of duplex conductor at least 60 feet (18.5 m) long. Measure the length accurately with a tape measure.

2 Connect the meter's red lead to one of the two pieces of wire and the black lead to the other (strip back the insulation on both leads first). (Don't twist the conductors together at the far end or you'll create a short circuit.)

3 Turn on the Fault Mapper, push the meter's test button, and note the indicated length.

4 Use the scroll keys to adjust the VOP value up and down until the test result equals the actual length of the cable. Your final VOP value will be appropriate for all cables of the same gauge and type.

I've tested many cable types and gauges and have found that all the cable commonly installed in boats falls in the range of 60% to 65% VOP. If you're off by a percentage point or so, it's not the end of the world, but it will create a small inaccuracy in the meter's calculation of distance to the fault. Table 3-1 gives the VOP values for commonly used American Wire Gauge (AWG) sizes of boat cable, along with metric equivalents.

Using the TDR to Locate a Fault

Before you can begin testing, you must determine the size of the cable or wire you are testing. All approved wire and cabling has its size printed on its insulation jacket (in AWG in North America; in mm² for the rest of the world) as in the top photo.

Once you know the wire size, follow these steps:

TABLE 3-1	Velocity of Propagation Values for Typical DC Boat Wiring	
AWG	Metric Equivalent (mm²)	VOP (%)
8	8	63
10	5	64
12	3	62
14	2	61
16	1	61

This wire's size—4.0 mm²—is printed on its insulation jacket. This is roughly equivalent to 10 to 12 AWG.

Attach the meter's leads to the small fuse panel at the source of the test circuit: the red lead goes to the DC positive source at the fuse, and the black lead to the DC negative. (Anywhere on this bus will do, as it is a common return for all the circuits on the panel.) The TDR screen displays a VOP value of 61%, which corresponds to the circuit's 14 AWG wire. The meter's internal battery status (fully charged) is also shown graphically at the bottom of the display.

The TDR has identified an open circuit 21.6 feet (6.6 m) away. If you know how the wiring harness in question is routed through the boat, you can work your way right to the problem to make the necessary repair. Does it seem too good to be true? It might be, if you don't have a clear understanding of what else might be implied by this reading.

1 Determine the cable's VOP using Table 3-1 and enter it into the TDR.

2 Connect the two leads to both conductors at one end of a duplex pair. (As noted earlier, a TDR cannot trace a single conductor.)

3 Turn off the power to the circuit being traced. If you don't, the Fault Mapper TDR will give you an on-screen prompt that voltage is present.

4 Push the "test" button on the unit, and the unit will trace the circuit.

The photo sequence shows the testing capabilities, as well as some limitations, of the TDR.

What's the issue regarding lightbulbs? The filament in the bulb completes the circuit by connecting the positive and negative feed conductors. The same thing happens if you twist the bare ends of two conductors together. The TDR's signal bounces from one of the paired conductors to the other, and in both cases the bouncing stops where the circuit is completed,

The TDR can "map" only as far as the first junction—a short circuit—in a circuit (the first limitation). To continue tracing a circuit beyond a junction, you must find the junction (the TDR has told you how far away it is) and isolate the next leg of the circuit by disconnecting the branch in question. For example, if a cabin light circuit contains a tee to feed a light on the opposite side of the boat, as shown here, you will have to disconnect that feed wire from the main harness and trace each branch of the circuit separately.

The TDR indicates a short circuit 30.6 feet (9.3 m) away, but this is a red herring. Another limitation with a TDR is that lightbulbs show up as short circuits. In this case, there is, indeed, a (perfectly functional) navigation light 30.6 feet away from the TDR.

at the farthest point from the source of the signal. The unit can only tell you where the bouncing stops; it can't determine if it's an open circuit, a short circuit, or a complete circuit. Thus, you really need to know how a circuit is laid out and what components are installed on it to correctly interpret the TDR's results.

Using the TDR to Check Wire Size

In addition to finding faults in a circuit, you can use a TDR to determine if a circuit complies with ABYC standards for wire size. Here are the facts we know about the circuit example above:

- The navigation light being tested in photo 4 opposite is working.
- The 12 V circuit is wired with 14 AWG wire.
- The total length of the circuit from the panel to the light and back to the source is 61.2 feet (30.6 feet × 2).
- The lightbulb draws 2 amps.

Navigation lights are a critical circuit, and the ABYC (and U.S. Coast Guard) permits a maximum 3% voltage drop. We then refer to the appropriate voltage drop table in ABYC Standard E-11, reproduced below.

NOTE: *In the event of a conflict between the voltage drop table and the ampacity table, use the larger wire size.*

Length of Conductor from Source of Current to Device and Back to Source - Feet

TOTAL CURRENT ON CIRCUIT IN AMPS.	10	15	20	25	30	40	50	60	70	80	90	100	110	120	130	140	150	160	170
12 Volts - 3% Drop Wire Sizes (gauge) - Based on Minimum CM Area																			
5	18	16	14	12	12	10	10	10	8	8	8	6	6	6	6	6	6	6	6
10	14	12	10	10	10	8	6	6	6	6	4	4	4	4	2	2	2	2	2
15	12	10	10	8	8	6	6	6	4	4	2	2	2	2	2	1	1	1	1
20	10	10	8	6	6	6	4	4	2	2	2	2	1	1	1	0	0	0	2/0
25	10	8	6	6	6	4	4	2	2	2	1	1	0	0	0	2/0	2/0	2/0	3/0
30	10	8	6	6	4	4	2	2	1	1	0	0	0	2/0	2/0	3/0	3/0	3/0	3/0
40	8	6	6	4	4	2	2	1	0	0	2/0	2/0	3/0	3/0	3/0	4/0	4/0	4/0	4/0
50	6	6	4	4	2	2	1	0	2/0	2/0	3/0	3/0	4/0	4/0	4/0				
60	6	4	4	2	2	1	0	2/0	3/0	3/0	4/0	4/0	4/0						
70	6	4	2	2	1	0	2/0	3/0	3/0	4/0	4/0								
80	6	4	2	2	1	0	3/0	3/0	4/0	4/0									
90	4	2	2	1	0	2/0	3/0	4/0	4/0										
100	4	2	2	1	0	2/0	3/0	4/0											
24 Volts - 3% Drop Wire Sizes (gauge) - Based on Minimum CM Area																			
5	18	18	18	16	16	14	12	12	12	10	10	10	10	10	8	8	8	8	8
10	18	16	14	12	12	10	10	10	8	8	8	6	6	6	6	6	6	6	6
15	16	14	12	12	10	10	8	8	6	6	6	6	6	4	4	4	4	4	2
20	14	12	10	10	10	8	6	6	6	6	4	4	4	4	2	2	2	2	2
25	12	12	10	10	8	6	6	6	4	4	4	4	2	2	2	2	2	2	1
30	12	10	10	8	8	6	6	4	4	4	2	2	2	2	2	1	1	1	1
40	10	10	8	6	6	6	4	4	2	2	2	2	1	1	1	0	0	0	2/0
50	10	8	6	6	6	4	4	2	2	2	1	1	0	0	0	2/0	2/0	2/0	3/0
60	10	8	6	6	4	4	2	2	1	1	0	0	0	2/0	2/0	3/0	3/0	3/0	3/0
70	8	6	6	4	4	2	2	1	1	0	0	2/0	2/0	3/0	3/0	3/0	3/0	4/0	4/0
80	8	6	6	4	4	2	2	1	0	0	2/0	2/0	3/0	3/0	3/0	4/0	4/0	4/0	4/0
90	8	6	4	4	2	2	1	0	0	2/0	2/0	3/0	3/0	4/0	4/0	4/0	4/0	4/0	
100	6	6	4	4	2	2	1	0	2/0	2/0	3/0	3/0	4/0	4/0	4/0				

The 3% voltage drop table from ABYC Standard E-11. (Courtesy ABYC)

For safety's sake, the ABYC recommends always "rounding up" when using voltage drop tables. In this case, that means we have to round up from 2 amps (actual) to 5 amps, the lowest value on the table. We also have to round up the length, from 61.2 feet (actual) to 70 feet (the next higher value on the table). This table shows that we should use 8 AWG wire for both the DC positive feed to the light and the DC negative return.

If this seems excessive, we can use the ABYC's circular mil area formula (from Standard E-11) as an acceptable alternative to the table. Let's work the numbers. Our factors for the problem are:

$$CM \text{ (circular mil area)} = (K \times I \times L) \div E,$$
where:

K (constant for the resistivity of copper) = 10.75

I (load current in amperes) = 2

L (total circuit length in feet) = 61.2

E (acceptable voltage drop of 3% for 12 V) = 0.36

Therefore:

$$CM = (10.75 \times 2 \times 61.2) \div 0.36 = 3,655$$

Next, compare the calculated CM value to the circular mil table from ABYC E-11 shown opposite. You'll see that even when rounding up to the next nearest size from 3,655 CM, we hit the 14 AWG row at 3,702 CM. That's good news; the circuit wiring is compliant.

There is often a huge difference between the CM calculation method and using the tables in E-11, as the above example illustrates. In fact, I rarely use the tables, and almost always perform the calculations using actual values, with no rounding. It often results in a smaller wire size requirement that still meets standards. For a boatowner, this can save several dollars on a single circuit. For professional technicians, it can save tons of copper per year!

Does the TDR play a major role in this? You bet. Because you can check the actual length of the wire runs, you can confirm compliance in factory-installed, aftermarket, and repaired wiring. No one should be able to pull a fast one on you by fudging wire sizing.

VALUE AND UTILITY

The tools described in this chapter are versatile and can be used for any type of wiring or circuit, AC or DC. (Just be sure the power is *off* when using them, since these tools can't function with power running through the wiring.) The tone-generating circuit tracer is a highly affordable tool. And while a bit more expensive, the AEMC Fault Mapper is also a tone generator, so instead of purchasing both tools, you can add the AEMC model TR02 receiver to the TDR and have the benefits of both methodologies described in this chapter combined into one kit.

In the case of the TDR, you may need to do a little additional homework to gather more VOP values, but if you're a marine electrician, you have the wire lying around on rolls. All you'll need are minimum 60-foot lengths to start building your own library of VOP values. Once you've done that, yet another value of the Fault Mapper will stand out—the ability to check inventory by checking how much cable is left on the roll! I'll never use an ohmmeter for this type of work again.

CONDUCTOR GAUGE	MINIMUM ACCEPTABLE CM AREA AWG	MINIMUM ACCEPTABLE CM AREA SAE	MINIMUM NUMBER OF STRANDS		
			TYPE 1*	TYPE 2**	TYPE 3***
18	1,620	1,537	-	16	-
16	2,580	2,336	-	19	26
14	4,110	3,702	-	19	41
12	6,530	5,833	-	19	65
10	10,380	9,343	-	19	105
8	16,510	14,810	-	19	168
6	26,240	24,538	-	37	266
4	41,740	37,360	-	49	420
2	66,360	62,450	-	127	665
1	83,690	77,790	-	127	836
0	105,600	98,980	-	127	1064
00	133,100	125,100	-	127	1323
000	167,800	158,600	-	259	1666
0000	211,600	205,500	-	418	2107

*Type 1 - Solid conductor and stranding less than that indicated under Type 2 shall not be used

**Type 2 - Conductors with at least Type 2 stranding shall be used for general purpose boat wiring.

***Type 3 - Conductors with Type 3 stranding shall be used for any wiring where frequent flexing is involved in normal use.

NOTE: 1.   Metric wire sizes may be used if of equivalent circular mil area. If the circular mil area of the metric conductor is less than that listed, the wire ampacity shall be corrected based on the ratio of the circular mil areas. For comparison of conductor cross sections (AWG and ISO) (See AP TABLE 2)

2.   The circular mil area given is equal to the mathematical square of the diameter of the AWG standard solid copper conductor measured in one thousandths of an inch.

$$\text{The area in square inches} = \frac{pi(circular\ mils)}{4(1,000,000)}$$

The circular mil area of the stranded conductors may differ from the tabulated values and is the sum of the circular mil areas of the wires (strands) in the conductor.

The conductor circular mil area and stranding table from ABYC Standard E-11. (Courtesy ABYC)

Testing AC and Grounding Systems for Polarity, Voltage Drop, Impedance, and Frequency

I n the "old days," the tools of choice for working around AC systems on boats were the DVOM and maybe an oscilloscope. Today, the advanced technician or boatowner can easily test conductor impedance, polarity, frequency level, voltage drop, and peak RMS voltage with a circuit analyzer.

THE SURETEST CIRCUIT ANALYZER

As far as I can determine, the SureTest circuit analyzer from Ideal Industries is a unique instrument that, although now in its second generation, still represents the only example of this new category of test devices. Our discussion will use model 61-164. (A sister model is available that adds a testing capability for arc fault circuit interrupters, but since AFCIs are not used on boats, this feature is not of interest to us.)

As with many of the tools outlined in this book, the SureTest 61-164 circuit analyzer uses a microprocessor, which allows it to perform six extremely important tests on an AC system in a matter of seconds. It comes with probes for testing receptacles and has an optional alligator clip. At around $250, I consider this tool to be worth every penny and an absolute necessity for anyone working around AC systems.

Now, let's go through each test function one by one.

> ### CIRCUIT ANALYZER

SureTest model 61-164, Ideal Industries, Inc., www.idealindustries.com

The SureTest 61-164 circuit analyzer.

Checking for Polarity

One of the most common errors in AC wiring is improper connection of the hot, neutral, or ground wire at receptacles, which can be an elusive problem. According to ABYC Standard E-11, most 120 V wired boats are required to have a reverse-polarity indicator at the main AC panelboard. However, the panel indicator only monitors from the shore-power pedestal to the panel; receptacles supplied from the panel are not monitored by the indicator. If they have been mistakenly miswired, the panel indicator won't pick it up, creating a potentially dangerous situation at that receptacle. By using the SureTest at every receptacle

on the boat, you can easily check the polarity and determine if you have to take action.

In this application, the SureTest replicates the functions of a $5 circuit tester that you probably already own. Don't worry; this instrument is capable of much more sophisticated and useful testing, but these basic tests are important, so we'll review them first.

Here are the basics for testing polarity:

1 Plug the SureTest into an active receptacle.

2 The tester's screen display should immediately light up.

3 The unit will progress from its opening screen to a display of three green LEDs indicating hot, ground, and neutral.

4 If all three LEDs are lit, it's a firm indication that polarity is correct—in other words, hot and neutral have not been reversed.

5 Other LED combinations indicate various AC wiring problems—an open-circuited (not connected) hot or neutral, no-ground situations, and false grounds.

The SureTest indicates different AC wiring problems by displaying various combinations of its three LEDs.

One of the limitations of the tester is that it will not detect reversal of the neutral and grounding conductors, which is a possibility if the receptacle installer was really careless. However, the grounding terminal on receptacles is always indicated by a green lug, so you can visually see the error. The grounding terminal is not obvious with the hot and neutral, and the SureTest will pick it up in a flash as a reversed polarity condition.

"False" Grounds

Since false grounds can create dangerous situations on a boat and/or in the water around it, they warrant further discussion. A false ground, for our purposes, is a ground connection to neutral at a location other than the source of power. In other words, it is a point where an AC neutral conductor is "falsely" connected to the AC grounding, or earth, conductor.

The U.S. National Electric Code (NEC) only allows a neutral-to-ground link at the distribution panel for AC systems. The ABYC, on the other hand, only allows neutral-to-ground links at sources of AC power on board the boat—primarily generators, inverters when they are producing AC, and the secondary side of isolation transformer sets, as shown in the illustrations. If the boat is not equipped with any of these devices, the neutral and ground must be separated all the way back to the shore-power distribution panel. At that point, the NEC code takes precedence.

Neutral-to-ground bonds are important at AC power sources because if there is a short circuit at an AC appliance, it means that the "hot" has shorted to the equipment case, effectively bypassing the neutral conductor. By tying the neutral and ground conductor together at the source of power, the fault current is provided with an alternative path back

Proper neutral-to-ground bonding of an isolation transformer, according to Standard E-11. (Courtesy ABYC).

Power Inlet (Electrically
insulated from the boat
if isolator is installed)

Transfer Switch
INV-OFF-SHORE
31.6.5.2

Branch Circuit
Breaker (Typical)

Shore Power
Cable
Connector

Shore
Connection

Shore Power
Cable

Main Shore Power
Disconnect
Circuit Breaker

120 VAC
Grounding
Type Recept.

Ungrounded Conductor (Black)
Grounded Conductor (White)
Grounding Conductor (Green)

2 Pole, 3 Wire
Grounding Type
Plugs & Receptacles

Optional
Galvanic
Isolator

Shore Side
Boat Side

Green
White
Black

Breaker
Required IF Not
Integral

Reverse Polarity Indicator

Black
White
Green

G   N   L
Inverter

31.5.3.6

Battery
Switch

See ABYC E-11

+

Battery

−

Main AC Grounding Bus

DC Grounding Conductor (31.6.3)

DC Negative

Engine Negative
Terminal / Bus

See ABYC E-11

Proper neutral-to-ground bonding of an inverter, according to Standard E-11. (Courtesy ABYC).

to the source and a circuit breaker will trip, shutting off the circuit. Remember that the green ground wire is attached to the case of AC equipment, which is key to the success of this configuration.

So here's the rub, but in the case of boats it's actually a useful one. The SureTest will identify these "false" grounds within about 15 to 20 feet of the occurrence of a neutral-to-ground link, including a distribution panel, onboard generator, inverter, or isolation transformer. So in a nutshell, "false" may be exactly what you are looking for at these devices. It's at AC appliances that you don't want this link. A good example of where a false ground may appear is at a 240 V electric range or clothes dryer that is wired for older residential applications, which are often installed on older boats (see photo page 50).

These neutral-to-ground links must be disconnected when installed on a boat (see photo page 51). Otherwise, every time you turn on the appliance, you run the risk of AC power (which would normally be returned down the neutral conductor) being pushed down the grounding conductor, and possibly entering the water via the DC bonding system. Remember that on ABYC-compliant boats, the AC and DC grounding systems should be linked either at the engine's negative terminal or, occasionally, at the grounding buses behind the boat's main AC and DC distribution panels.

Proper neutral-to-ground bonding of a 120 VAC generator, according to Standard E-11. (Courtesy ABYC).

A neutral-to-ground link on the back of a 120/240 V electric range—a false ground. Another common appliance where this may occur is electric clothes dryers.

So the SureTest's indication of a "false" ground may be either a good or a bad thing, depending upon whether it has identified a proper link at an AC power source or an improper one at an AC appliance or a receptacle. You'll need to confirm or deny which is the case by performing a little visual inspection detective work of all the devices mentioned, and physically tracking down the neutral-to-ground link.

Voltage Measurements

The second test in the sequence is the line voltage measurement. Categorically, line voltage should be maintained at a maximum of ±10% of the rated voltage. For 120 V potential, voltage should be between 108 and 132 VAC; for 220 systems, the range is 198 to 242 VAC.

⚠️WARNING

APPLIANCE GROUNDED TO NEUTRAL CON-
DUCTOR THROUGH A LINK. IF USED IN A MOBILE
HOME OR IF LOCAL CODES DO NOT PERMIT
GROUNDING THROUGH THE NEUTRAL, DIS-
CONNECT THE LINK FROM THE NEUTRAL AND
USE GROUNDING TERMINAL TO GROUND
APPLIANCE IN ACCORDANCE WITH LOCAL CODES.
CONNECT NEUTRAL TERMINAL TO BRANCH
CIRCUIT NEUTRAL IN USUAL MANNER. IF
APPLIANCE IS TO BE CONNECTED WITH A
CORD KIT, USE 4-CONDUCTOR CORD.

CONNECT TO INDIVIDUAL BRANCH CIRCUIT WITH
MINIMUM SUPPLY-CIRCUIT CONDUCTOR AMPACITY
OF 30 AMPS AND 30 AMP MAXIMUM NON-TIME
DELAY FUSE, DELAY FUSE OR CIRCUIT BREAKER.
THIS APPLIANCE MUST BE GROUNDED ACCORDING
TO THE REQUIREMENTS OF NATIONAL AND LOCAL
CODES IN ORDER TO AVOID THE POSSIBILITY OF
HAZARD DUE TO ELECTRICAL SHOCK.

Always look for labeling on the back of an appliance for grounding instructions, and be absolutely certain that the neutral-to-ground link is disconnected on a boat installation.

Since the SureTest 61-164 will also measure peak voltage, you can get a feel for distortion by calculating the theoretical peak at line voltage × 1.414, and comparing it to the actual measured peak. You won't be able to see the distortion as you can with the oscilloscope, but you'll be able to determine if it exists, and calculate it if it is excessive (see Chapter 7).

Besides basic line voltage and peak values, the unit will also test for neutral-to-ground leakage. (Press the "advance" arrow key to go to the next test in the sequence.) This value should always be less than 2 VAC. More than 2 VAC is an indication of excessive current leakage from neutral to ground. This condition is typically caused by the breakdown of insulation within an appliance, or possibly by two terminals that are barely

touching, creating a partial, though high-resistance, connection. In a 220 V multiple-phase system, high neutral-to-ground leakage could indicate severely unbalanced legs between the phases or excessive harmonic distortion on the shared neutral. These conditions can create equipment performance problems, as described in Chapter 7.

Table 4-1 on page 52 summarizes voltage readings, expected results, problems, and possible causes and solutions.

Voltage Drop Measurements

As already stated, excessive voltage drop is one of the marine electrical system's biggest enemies. I've also noted that the primary by-product of excessive electrical resistance is heat, and that too much voltage drop is the same as excessive resistance. This is an area where the SureTest really shines. As far as I know, it's the only test instrument that can apply a load to the wiring at three levels (12 amps, 15 amps, and 20 amps) and calculate the voltage drop as a percentage within a few seconds.

With this information in hand, troubleshooting is easy. Based on NEC recommendations, anything in excess of 5% is too much. If excessive voltage drop exists, it may mean that the wire gauge is too small for the task at hand, or there is a poor-quality connection somewhere in the circuit. To confirm or reject this, simply retest at points closer to the source and see if the problem goes away as you get closer to the source. If it does, then the problem is between the high-reading checkpoint and the low-reading checkpoint, with several possible causes:

- a loose connection in that leg of the circuit
- undersized cabling, perhaps added to the original circuit
- a faulty receptacle that needs replacing

TABLE 4-1	Voltage Troubleshooting Tips			
Measurement	**Expected Result**	**Problem**	**Possible Causes**	**Possible Solutions**
Line voltage 120 VAC	108–132 VAC	High/low	Too much load on the circuit.	Redistribute loads on the circuit.
Line voltage 220 AC	198–242 VAC		High-resistance connection within the circuit or at the panel.	Locate high-resistance connection/device and repair/replace.
			Supply voltage too high/low.	Consult power company.
Neutral-ground voltage	< 2 VAC Voltage	High G–N > 2 VAC	Current leaking from neutral to ground.	Identify source of leakage: multiple bonding points, equipment, or devices.
			Unbalanced three-phase system.	Check load balance and redistribute load.
			Triplen harmonics returning on neutral in three-phase system.	Oversize neutral to impedance. Reduce harmonic effect via filter or other methods.
Peak voltage 120 VAC	153–185 VAC	High/low peak voltage	Supply voltage too high/low.	Consult power company.
Peak voltage 220 VAC	280–242 VAC		High peak loads on line caused by electronic equipment on line.	Evaluate number of electronic devices on circuit and redistribute if necessary.
Frequency	60 Hz	High/low frequency	Supply frequency too high/low.	Consult power company.

Source: Ideal Industries, Inc.

Physically check these possibilities and repair or replace cables or components as needed.

This is also an area where the infrared heat gun (see Chapter 5) can be useful in finding the hot spot in the circuitry. Use Table 4-2 to help you troubleshoot voltage drop problems.

Impedance Measurements

To further assist you in troubleshooting excessive voltage drop, the SureTest can measure the exact impedance (resistance) in each of the three conductors in a circuit. Again, I know of no other tool that has this capability.

If measured voltage drop is more than 5%, a problem exists. Using the meter's arrow keys, scroll through the menu to read and compare the impedance values for the hot, neutral, and ground conductors. What you are looking for here is parity among the conductors.

If impedance in one conductor is significantly higher than another, then you've narrowed down the problem. Again, loose or corroded connections are the most probable cause, and you must now physically search out the culprit.

If all the conductors read high, then the cabling is probably undersized. Keep in mind

TABLE 4-2	Voltage Drop Troubleshooting Tips			
Measurement	**Expected Result**	**Problem**	**Possible Causes**	**Possible Solutions**
Voltage drop	< 5%	High voltage drop	Too much load on the circuit.	Redistribute the load on the circuit.
			Undersized wire for length of run.	Check code requirements and rewire if necessary.
			High-resistance connection within the circuit or at the panel.	Locate high-resistance connection/device and repair/replace.

Source: Ideal Industries, Inc.

that if the circuit is protected by a GFCI (ground fault circuit interrupter) or a GFP (ground fault protection) device, the device will trip during this test. Simply reset it after you've acquired the values.

Use Table 4-3 to help you troubleshoot impedance problems.

Measuring Available Short-Circuit Current

The SureTest can also measure (in kiloamps, kA) the amount of available short-circuit current (ASCC)—how much current can flow through a circuit in spite of resistance factors such as wire gauge and terminal connections. This will be

TABLE 4-3	Impedance Troubleshooting Tips			
Measurement	**Expected Result**	**Problem**	**Possible Causes**	**Possible Solutions**
Hot and neutral impedance	< 0.048 Ω/foot of 14 AWG wire < 0.03 Ω/foot of 12 AWG wire < 0.01 Ω/foot of 10 AWG wire	High conductor impedance	Too much load on branch circuit.	Redistribute the load on the circuit.
			Undersized wire for length of run.	Check code requirements and rewire if necessary.
			High-resistance connection within the circuit or at the panel.	Locate high-resistance connection/device and repair/replace.
Ground impedance	< 1 Ω to protect people < 0.25 Ω to protect equipment	High ground impedance	Undersized wire for length of run.	Check code requirements and rewire if necessary.
			High-resistance connection within the circuit or at the panel.	Locate high-resistance connection/device and repair/replace.

Source: Ideal Industries, Inc.

significant when you consider circuit breaker ratings, because if the ASCC is greater than the ampere interrupting capacity (AIC) rating for a breaker or fuse, the device may not function when a short circuit occurs. Instead, in the case of the breaker, the contacts can be welded closed, and in the case of a fuse, the fuse element can arc.

ABYC Standard E-11 establishes AIC limits for AC and DC breakers and fuses, as shown in the ABYC table below. Simply put, whatever readings the SureTest gives you on this test sequence, you need to be sure they are below the AIC ratings shown in the table. Typical AIC ratings for circuit breakers used in AC applications will exceed this value, in my experience. For example, on docks I've checked, the ASCC readings were well below the AIC ratings on the breakers used due to the voltage drop induced by the wire gauge and length of wire runs typical of marina and dock wiring schemes. Common ASCC readings on docks I've checked fall in the 1,000 to 2,000 amp range.

GFCI and GFP Testing

One more useful SureTest function is testing GFCI and GFP devices. (GFPs are also known as RCDs—residual current devices, and EPDs—equipment protecting devices.) The ABYC describes GFCI devices as those intended to protect people from shock hazards associated with AC faults, and GFP devices as those intended to protect equipment from damage. The big difference in this application between a SureTest and a $5 LED circuit tester is that a SureTest will measure both the trip rate (in milliamps, mA) and the trip time for the GFCI or GFP. This information can warn you of the device's imminent failure. Considering the statistically high failure rate of GFCIs, particularly in marine applications (more on GFP testing in a minute), the ability to perform these tests is of paramount importance.

During the test, the instrument creates a small but steadily increasing current imbalance between the hot and neutral conductors. The GFCI should respond by tripping between 4 mA and 7 mA. Most GFCIs that I've checked trip at 7 mA if all is well. Note that the nominal rating of GFCIs in North America is 5 mA, but as with all breaker-type devices, the nominal rating and the actual trip rate are always different values.

SHORE POWER SOURCE	MAIN SHORE POWER DISCONNECT CIRCUIT BREAKER	BRANCH BREAKER
120V – 30A	3000	3000
120V – 50A	3000	3000
120/240V – 50A	5000	3000
240V – 50A	5000	3000
120/208V – 3 phase/WYE – 30A	5000	3000
120/240V – 100A	5000	3000
120/208V – 3 phase/WYE – 100A	5000	3000

NOTES: 1. *The main circuit breaker shall be considered to be the first circuit breaker connected to a source of AC power. All subsequent breakers, including sub-main breakers connected in series with a main circuit breaker, shall be considered to be branch circuit breakers.*
2. *A fuse in series with, and ahead of, a circuit breaker may be required by the circuit breaker manufacturer to achieve the interrupting capacity in TABLE V-B.*

The circuit breaker interrupting capacity for systems over 50 V from ABYC Standard E-11. (Courtesy ABYC)

TABLE 4-4	GFCI Troubleshooting Tips			
Measurement	**Expected Result**	**Problem**	**Possible Causes**	**Possible Solutions**
GFCI test	GFCI trips within trip time.	GFCI doesn't trip within proper trip time.	GFCI may be installed improperly.	Check wiring for proper installation in accordance with manufacturer's instructions and NEC.
		GFCI doesn't trip.	GFCI may be defective.	Check wiring and ground. Replace GFCI if necessary.

Source: Ideal Industries, Inc.

Here's how to perform the GFCI test:

1 Plug the tester into the GFCI or GFCI-serviced receptacle.

2 Press the "GFCI" button on the tester.

3 If the GFCI is working, it will trip, and the instrument will display the trip rate in mA just before it blinks out.

4 To display the trip time, simply reset the GFCI. How much time is too much? The nominal trip delay time for GFCI devices is 0.025 second or 25 milliseconds (ms).

5 If your readings are not close to these values, replace the GFCI receptacle.

To test GFPs, simply depress the side arrow key on the face of the tester to the 30 mA setting and perform the test as described above. This is an extremely valuable capability because it corresponds to the 30 mA trip rate that is a requirement for whole-boat protection everywhere in the world except North America. Such whole-boat protection consists of a 30 mA rated GFP installed at the boat's main shore-power connection. Any current leakage on the boat that gets to the 30 mA level will turn off the power to the entire boat. (I am lobbying at the standards development level for the United States to adopt this requirement.) Typically, boats coming to the United States from Europe, New Zealand, and Australia are equipped with these devices, and any U.S. boats shipped to those regions must install them. Boats from Asia, on the other hand, are often built specifically for the U.S. market and are therefore wired to North American standards.

Use Table 4-4 to help you troubleshoot GFCI problems.

The following photos show the complete test sequence of this very useful device.

Circuit testing, similar to what you can do with a simple $5 LED tester. In this case, the SureTest indicates "no ground," a dangerous condition.

Testing line voltage. The SureTest shows 119 V at the outlet. Any reading within ±10% of nominal voltage (120 V) is acceptable.

Testing neutral-to-ground leakage (voltage). The SureTest shows 0.9 V, which is acceptable.

Testing peak voltage value. The SureTest shows 168.5 V, which is normal.

Testing frequency in Hz. The SureTest shows 60 Hz, which is the expected result; ±5% is the maximum accepted tolerance.

Measuring voltage drop with a 15 amp load. The SureTest shows 15.5%, which is too high; it should be 5% or less. High readings like this can indicate poor-quality terminations in the system or undersized wiring.

Checking voltage drop with a 20 amp load. The SureTest shows 20.7%, which again is too high.

Testing ASCC. The SureTest shows 0.08 kA (80 amps). This result won't cause problems for the circuit breaker, but it does indicate that undersized wiring may have been used to construct the circuit. (The smaller the wire, the less current it can carry.)

Measuring the actual (as opposed to the rated) trip rate for a GFCI device. The 7.3 mA value is perfectly normal for a 5 mA rated device.

Measuring the trip rate for a GFP. The SureTest shows 30.6 mA, which is close to the 30 mA trip rate required for whole-boat protection (outside the United States).

A homemade 30 to 15 amp adapter for checking dock wiring.

Other Options

The SureTest comes equipped with enough probes for testing receptacles, while the optional 61-183 alligator clip adapter is a useful option for checking power supplies that are hard-wired to AC equipment. You will also need to make an adapter (see the sidebar on page 58) for the shore-power connection, like the one shown above, and possibly several adapters if you plan to test various services (30 amps, 50 amps, etc.).

For 50 amp, 240 V service, I usually check L1 and L2 as two separate entities using two 50 amp plug sets that key into the L1 hot and L2 hot terminals. Keep in mind that in North America, a 240 V system is similar to two

MAKING SURETEST ADAPTERS

It's relatively easy to make adapters to use the SureTest to check 240 V, 50 amp shore-power service. You will need to buy two 50 amp replacement plug assemblies (male plugs) and two 15 amp female plug assemblies to plug the SureTest into. You'll also need about 2 feet of four-wire 240 V, 50 amp shore cord or the equivalent.

You need to make two assemblies so you can test the ground, neutral, and 120 V "hot" lead for each leg of the 240 V system. Construct one assembly (B) so that the hot lead connects to the L2 terminal (see the illustration) and the grounds (green or GR) and the W (white) terminals connect to the 240 V plug. Construct the other adapter (A) the same way, except connect the hot lead from the 15 amp plug's small-bladed terminal to the 240 amp plug's L1 terminal.

Use about 1 foot of the shore-power cabling for each adapter. Mark the L1 and L2 connections with an indelible marker for future reference. The illustration shows how the completed adapters should look.

50 amp, 240 V plug assemblies

standard 15 amp plug assemblies

black, L1

white

gray, not connected

A

green

gray = neutral

black = "hot"

gray, L1, not connected

white = neutral

black, L2, connected

B

green

gray

black

The L1 and L2 conductors and the neutrals and grounds. In A, L1 in the 50 amp plug is activated from the 15 amp assembly. Both white terminals are activated, and both green terminals are activated as grounds. In B, the black terminal in the 50 amp plug is the L-shaped prong, which becomes L2, and is connected to the black terminal on the 15 amp plug. The white and green terminals are continuous.

120 V systems combined, so each L, either 1 or 2, is actually a separate 120 V service. Remember the tool only comes equipped with a three-prong 15 amp plug set, so you will have to make some plug modifications for marine applications.

VALUE AND UTILITY

The SureTest is the tool I use most with AC circuit analysis, and I think anyone working with AC systems should own one. When you use it, keep in mind that the electrical tolerance specifications discussed above apply to most of North America, and that other areas (that provide 120 V, 60 Hz service or 240 V, four-wire service) may have different requirements, which you should confirm locally.

Temperature Monitoring as a Diagnostic Aid

The primary by-product of excessive electrical resistance is heat, which can be an electrical system's worst enemy. Along with causing early failures of expensive electrical and electronic equipment, excessive heat can cause fires. A recent study conducted by BoatU.S. determined that 55% of all boat fires are caused by faulty electrical installations. It's no wonder the ABYC addresses excessive electrical resistance and heat buildup in several standards dealing with electrical installations and equipment.

Since heat is such an important issue, a few years ago I decided that I should measure it in electrical systems, just as I had always done with engine cooling systems and refrigeration systems. Enter the heat sensor or heat-sensing gun, which is basically a handheld infrared thermometer. The unit uses an invisible infrared beam to sense the energy that is emitted, transmitted, or reflected from an object and translates that data into temperature. It also has a laser beam for aiming and an internal battery. They range in price from $30 to $150.

Because the laser and infrared beams can go where you cannot go, or perhaps may not want to go (remember we're dealing with excessive heat), it is a handy device for locating a number of potential problems:

- loose cable termination points
- damaged or corroded terminals
- undersized wiring
- faulty relays and switches (these will run hotter than normal if they have bad internal connections)

INFRARED HEAT SENSORS

Model 42500, Extech Instruments, www.extech.com (replaces model 42520)

Model TG-600, Greenlee, www.greenlee.com

60 series, Fluke, www.fluke.com

Model T-7350, Professional Equipment, www.professionalequipment.com

Model 22-325, RadioShack, www.radioshack.com

units have a range from slightly below 0°F to 500°F or so (−18°C to 260°C), which is fine for electrical troubleshooting and work on air-conditioning and refrigeration systems. However, if you intend to use the gun for other areas of troubleshooting as well, such as looking for temperature variations between cylinders on engines or tracing coolant flow through the engine's cooling system, you may want to spend a little more and get a unit that can read up to 1,000°F (538°C) or even higher. (Typical operating temperatures of cylinders and exhaust systems can exceed 1,000°F.) It's also useful if the unit displays results both in degrees Fahrenheit and Celsius (outside the United States, Celsius is the temperature-measuring unit of choice, and equipment manufacturers may list performance specifications in either units).

The RadioShack model listed in the sidebar will work fine for electrical troubleshooting, but its measurement scale (0°F to 400°F/−18°C to 204°C) may not be sufficient for the entire range of applications discussed here. For engine work, exhaust systems, and other high-temperature applications, look for a unit with a range of approximately 5°F to 1,000°F (−15°C to 538°C).

The Extech #42520 infrared heat-sensing gun. (Note: This model was recently replaced by model #42500.)

INFRARED HEAT-SENSING GUN

So what should you look for when choosing a heat-sensing gun? Two specifications are important: temperature range, and distance-to-spot ratio.

Temperature Range

The most important specification is the heat gun's temperature range. Some inexpensive

Distance-to-Spot Ratio

The second specification to consider is the device's distance-to-spot ratio, which is the ratio of the distance to the object (D) to the diameter of the infrared beam (S), or D:S. Simply put, as the infrared beam moves farther away from the gun, the diameter of the beam widens. Consequently, as the effective width of the infrared beam increases, the device measures temperatures across a larger area. For example, with a 12:1 ratio, if you aim at a spot 24 inches away, the diameter of the beam at the spot will be roughly 2 inches; with a 6:1 ratio, if the meter is 12 inches from the target, the diameter will be 2 inches. The illustration shows two ratios,

25 mm @ 50 mm @ 100 mm @
150 mm   300 mm   600 mm

$$\frac{D}{S} = \frac{6}{1}$$

S

infrared
heat sensor gun

1"   2"   4"
@ 6"   @ 12"   @ 24"

D

D = distance   S = spot size

area of coverage

29 mm @ 38 mm @ 57 mm @
150 mm   300 mm   600 mm

$$\frac{D}{S} = \frac{12}{1}$$

S

infrared
heat sensor gun

1.1"   1.5"   2.3"
@ 6"   @ 12"   @ 24"

D

D = distance   S = spot size

area of coverage

Infrared distance-to-spot ratios for both a 6:1 and 12:1 configuration.

6:1 and 12:1, with various distances. (Note: The ratio is calculated in millimeters, so the standard inch measurements only approximate the ratio.)

Generally you will want to measure the temperature of a small target, such as an electrical termination, so a larger distance-to-spot ratio is better, providing you with the flexibility to take accurate measurements from a greater distance.

In practice, I have found a 6:1 ratio to be satisfactory for boat work. (The Extech unit shown in this chapter has a 6:1 ratio, which limits its useful distance to a maximum of about 4 feet/1.2 m; otherwise the sample area gets too large.) I've found this distance to be quite satisfactory for anything I've ever needed around boats. As I'm typically checking the temperatures at wire and cable termination points or along the length of a conductor, I don't need a long range, nor do I want to cover a very large area in my measured sample.

Using an Infrared Heat Gun

Learning to operate an infrared heat-sensing gun is relatively easy. However, knowing how to use it and interpret the results is a bit more complicated. Here are some important points to keep in mind:

- As noted earlier, electrical resistance produces heat. An electrical problem often results in increased resistance, and thus increased heat. So we are looking for abnormal heat readings, not necessarily high temperatures, because as we all know, some electrical devices normally run hot during operation.

- A change in temperature (for example, from one point in the circuit to another) is more important than the temperature reading on the heat gun. In other words, we are looking for a temperature differential. (The actual average temperature will vary depending upon ambient temperature changes in the area you are working in. The trick is to look for abnormalities.)

- If there is no current flow, there will be no temperature change. The circuit in question needs to at least be trying to work for the heat gun to acquire any relevant temperature readings.

It's best to use the gun after everything has been shut down for a bit, and the ambient temperature has stabilized throughout whatever system you are checking. The reason is that after a system has been in use for a time, thermal transfer from one part to another has occurred (i.e., the temperature has risen in the space), and the pronounced temperature differential you are looking for will be much harder to find. Copper wire is not only a good conductor of electricity, it's also a great conductor of heat. (In fact, it can act as sort of a heat sink for whatever it's attached to.) If a component and its wire are at the ambient temperature, and the connection is loose, then almost immediately after you turn on the device, you'll see a marked increase over ambient temperature at the connection.

I have found that the best way to use a heat gun is to first scan the area where you will be working to determine the average ambient temperature. This is your benchmark. Then, when you scan a specific termination or device and find a much higher temperature (e.g., 25°F higher), you will know you've discovered a potential problem.

Let's look at a common real-world application: I have tested my boat's charging system and gotten some lower than normal readings, which could be caused by a loose connection anywhere in the alternator/battery circuit. I start checking for a bad connection at the positive battery terminal on the back of an alternator. I run the test with the 225 hp Yamaha outboard engine running, and the alternator is working pretty hard charging up

Here the heat gun is testing the quality of the main battery connection at the starter motor on my engine. The thermometer shows a reading of 90°F (32°C), which is close to the outside temperature. If the temperature at the connection rises significantly when cranking the engine, a faulty connection is indicated. In this case, all was well and no change in temperature was indicated while cranking.

some house bank batteries that have been discharged.

I point my heat gun at the front of the alternator and get a average reading of 186°F (85.6°C). As I move my laser beam around a bit, I point it directly at the stator winding through a ventilation opening near the middle of the alternator case. I get a reading of 211°F (99.4°C)! As I move the laser beam to the back of the alternator and point it directly at the positive battery terminal, I get a reading of 196°F (91°C). As I track my laser beam down the conductor, away from the alternator, the temperature gradually drops to 163°F (72.8°C).

What have I learned? Nothing, other than the fact that the alternator gets pretty darn hot! The wire is conducting some of the heat generated by the alternator at the terminal, eventually reaching the average temperature in the small engine area of the boat.

No problem was indicated by this test, but let's run it again, starting with a stone-cold engine. As soon as I turn on the engine, I add some electrical loads—all the onboard DC lighting, the blower fan, and navigation equipment—to get the alternator working hard. Now I quickly take some measurements, before the engine, components, and the compartment have a chance to heat up.

This time, I see a rapid increase of 25°F at the terminal compared to the surrounding area. Because this large temperature differential is so isolated, it's a pretty good indication that the terminal should be removed and inspected for damage or corrosion. Remember that the amount of heat produced varies directly with the amperage in the circuit and the resistance in the conductor or termination point (more amps = more heat potential).

Note that only high-current DC circuits—those connected to things like bow thrusters, starter motors, alternators, and anchor windlasses—have the capability to generate enough heat for the infrared heat gun to be a useful diagnostic tool. On the AC side of things, the gun is useful for tracing from a shore-power pedestal through to the boat and at receptacles that have loads plugged in and operating. The heat will invariably show up at or near the terminal ends, assuming that the wire gauge sizes were appropriate to begin with.

OTHER USES FOR TEMPERATURE MONITORING

Up to now, we have used temperature monitoring as a means for identifying electrical problems. Now let's shift our focus a bit to monitoring temperatures to avoid problems. Admittedly, this falls outside the normal scope of troubleshooting, so let's call this exercise "trouble prediction and avoidance" instead.

Batteries, Battery Chargers, and Inverters

Some components on boats need to be kept as cool as possible to ensure proper or even continued operation. Examples include isolation transformers, DC-to-AC inverters, and battery chargers, all of which generate fairly high temperatures.

In the case of battery chargers and inverters, the issue is important enough for the ABYC to address it in a standard. ABYC Standard A-31 (adopted July 2005) for manufacturers of battery chargers and inverters states, "Battery chargers and inverters shall be designed to operate at 122°F (50°C) continuously and be able to withstand a maximum of 158°F (70°C)."

Regarding batteries, we saw in Chapter 2 how temperature monitoring was one function of the Midtronics inTELLECT EXP-1000. In some marginal situations, the temperature of a battery can affect the test results, in which case the diagnostic meter will direct you to measure the battery temperature using its built-in infrared sensor (measurement

The infrared temperature sensor on the Midtronics inTELLECT EXP-1000 diagnostic meter.

The temperature display on the inTELLECT EXP-1000.

range from −20°F to 200°F) so it can compensate for the temperature.

In this same vein, many battery chargers have potentiometers that allow for temperature compensation adjustments based on the average temperature of the battery compartment.

Temperature compensation adjustment on a battery charger.

These adjustments, which maximize battery charging voltage output potentials, are often overlooked, but they can make a real difference in battery performance and longevity. The infrared heat-sensing gun can help here as well.

Galvanic Isolators

Temperature can also be a critical issue with diode-type galvanic isolators in several ways. First, a high case temperature may indicate a serious AC fault on the boat and a safety issue. There could be enough AC fault current flowing through the isolator to heat the case but not high enough current to trip a breaker, creating a shock hazard.

Second, galvanic isolators have been known to fail due to excessive ambient temperatures in their mounting locations. It is not at all uncommon for low-level currents to be running through an isolator for indefinite periods of time, resulting in a buildup of heat. This heat must be dissipated; otherwise, the diodes that are the key components within the isolator can be damaged from the thermal overload. ABYC Standard A-28 for manufacturers of galvanic isolators defines the ambient temperature for isolators that are either ignition-protected or intended for installation in machinery spaces at 122°F (50°C). For non-ignition-protected isolators and those that are not intended for installation in machine spaces, ABYC pegs the ambient temperature at 86°F (30°C).

VALUE AND UTILITY

Before installing a galvanic isolator, battery charger, or other similar device on your boat, it makes sense to check temperatures to determine the suitability of a specific location and/or the need for supplemental ventilation. (In fact, this *should be* part of a professional-grade installation.) A normal thermometer will do the job, but an infrared heat gun will generate results much faster.

Temperatures matter in electrical work, and excessive temperatures are a bad thing. Careful monitoring of equipment, circuits, and spaces will help you identify problems before they become serious or even dangerous. With an infrared heat sensor, you also can accurately determine when equipment needs servicing or replacing and have solid information for installation decisions.

Using Power Factor to Determine Energy Efficiency

Sooner or later, marine electricians and boatowners are confronted with the term *power factor*. For work on small boats, power factor is usually not too much of a concern, but as boats get bigger and bigger, with more AC-powered gear than ever installed, it warrants careful consideration and perhaps even monitoring.

Warning: At 110 V or 240 V, AC power on boats—whether shore power or provided by AC generators and DC-to-AC inverters—is potentially lethal. If you are not completely confident in your abilities around AC circuits, call in a specialist.

UNDERSTANDING POWER FACTOR

What is power factor? One definition states it as: "The ratio of real power to apparent power delivered in an AC electrical system or load. Its value is always in the range of 0.0 to 1.0 or 0% to 100%. A unity power factor (1.0) indicates that the current is in phase with the voltage and that reactive power is zero."

Hmmm. Everybody still with me? Let's break that definition down a bit:

- *Factor* describes a proportional relationship between two quantities—in this case, real power and apparent power.
- *Real power* (also known as *active power* or *working power*) describes the amount of electrical energy that is converted into useful work.
- *Apparent power* describes the total electrical energy actually delivered by the power supply (i.e., a utility company, an AC generator, or a DC-to-AC inverter).
- *Reactive power* is the difference between the power delivered and the power converted to useful work.

It may be helpful to think of power factor as the amount of current and voltage the customer actually uses compared to what the utility supplies. A high power factor is considered to be 90% or more.

Reactive Power

While reactive power may seem to be loss of power, it isn't a complete loss. In fact, reactive power is necessary to generate the magnetic field that is essential for the operation of motors, transformers, and similar devices with a wound coil. When a refrigerator compressor motor is turned on, there's a brief time lag between the moment the thermostat trips and the moment the motor begins to turn. It is during this ever-so-brief interval that reactive power is working. This is part of the reason why motors on start-up typically draw anywhere from three to six times their running current.

Reactive power needs to be controlled so that the devices function, but the public electric utility (or the boat's generator or inverter) doesn't recognize the inherent disparity, which in effect requires delivery of more power over a period of time. The amount of reactive power required needs to be minimized. All of this equates to more than simple inefficiency. Ultimately it affects the entire power-delivery system—things like transformer sizing, wire gauge appropriate to handle the increased current demands, and the like.

When you apply this concept to equipment and appliances, you move into the area of power consumption. How do your appliances and equipment rate? Low-power-factor items can have a significant impact on your boat's electrical system. For example, a low-power-factor fluorescent light ballast will draw more current than a high-power-factor ballast for the same wattage lamp. The result is that fewer AC appliances can operate on a given circuit.

On a boat with many AC circuits powering pumps and motors for refrigeration systems, air conditioners, and such, it's possible to have a fairly low cumulative power factor (i.e., the power factor for the total amount of watt-hours consumed). In extreme cases, this may require additional shore-power service to the boat.

IMPROVING YOUR POWER FACTOR

Power factor is most commonly improved by incorporating capacitor banks in a device. The capacitors charge and discharge on demand to supply the reactive power needed by the load. By engineering the capacitor bank correctly, an electrical designer can achieve, or at least get very close to, the desired 1.0 unity factor mentioned above. Once power factor correction circuitry is installed, the need to draw the reactive power from the AC supply source is eliminated or greatly minimized.

You can test the equipment and appliances on your boat to determine their power factor (see below). You may be able to upgrade to appliances and equipment that have a better corrected average power factor. All appliances that are "Energy Star" qualified have high power factors, which can significantly improve your power efficiency. For example, in the home construction field, a home that is certified as Energy Star qualified can be as much as 15% to 30% more electrically efficient, in terms of watt-hours used per month.

DETERMINING POWER CONSUMPTION

Performing an AC system load analysis is an important part of ensuring that there is enough available capacity to operate a boat's equipment based on typical use and consumption habits. (Note that the same applies to DC service, but we will be dealing with AC service only.) In many cases, boaters are attempting to add air conditioning or refrigeration systems to older boats. Such systems typically use quite a bit of power and often require installing additional, isolated shore-power service to the boat, which can be expensive. (Using an Energy Star appliance might make the difference between having to

AC LOAD ANALYSIS (FROM ABYC STANDARD E-11)

11.10.2. FOR AC SYSTEMS

11.10.2.1. Power Source Options

The method shown in E-11.10.2.2 shall be used for calculating the total electrical load requirements for determining the size of panelboards and their feeder conductors along with generator, inverter, and shore-power capacities. The total power required shall be supplied by one of the following means.

11.10.2.1.1. Single Shore-Power Cable. A shore-power cable, power inlet, wiring, and components with a minimum capacity to supply the total load as calculated, complying with E-11.7.2.1.1.

11.10.2.1.2. Multiple Shore-Power Cables. Multiple shore-power cables, power inlets, wiring, and components shall have a minimum total capacity to supply the total load as calculated complying with E-11.7.2.1.1. All sources need not be of equal capacity, but each power inlet shall be clearly marked to indicate voltage, ampacity, phase (if a three-phase system is incorporated), and the load or selector switch that it serves.

11.10.2.1.3. Onboard AC Generator(s) or Inverter(s). Onboard AC generator(s) or inverter(s) to supply the total load as calculated. Total minimum installed kVA [kilovolt amperes] for a single-phase system is as follows:

$$\text{kVA} = \text{Maximum Total Leg Amps} \times \text{System Voltage} \div 1000$$

11.10.2.1.4. Combination of Shore-Power Cable(s), Onboard Generator(s), and Inverter(s). A combination of power sources, used simultaneously if the boat circuitry is arranged such that the load connected to each source is isolated from the other in accordance with E-11.5.3.6. Shore-power cable(s) plus onboard generator(s) and inverter(s) capacity shall be at least as large as the total electrical load requirements as calculated. Generator(s) and inverters(s) installation and switching shall be as required in E-11.7.3.

11.10.2.2. Load Calculations

11.10.2.2.1. The following is the method for load calculation to determine the minimum size of panelboards and their main feeder conductors as well as the size of the power source(s) supplying these devices. (See E-11.10.2.1.)

11.10.2.2.1.1. Lighting Fixtures and Receptacles.

Length times width of living space (excludes spaces exclusively for machinery and open deck areas) times 20 watts per square meter (2 watts per square foot).

Formula:

$$\text{Length (meters)} \times \text{width (meters)} \times 20 = \underline{\hspace{2cm}} \text{ lighting watts, or}$$

$$\text{Length (feet)} \times \text{width (feet)} \times 2 = \underline{\hspace{2cm}} \text{ lighting watts}$$

11.10.2.2.2. Small Appliances, Galley and Dinette Areas.

Number of circuits times 1,500 watts for each 20-ampere appliance circuit.

Formula:

$$\text{Number of circuits} \times 1,500 = \underline{\hspace{2cm}} \text{ small appliance watts}$$

(continued)

11.10.2.2.3. Total.

Formula:

Lighting watts + small appliance watts = _____ total watts

11.10.2.2.4. Load Factor.

Formula:

First 2,000 total watts at 100% = _____

Remaining total watts × 35% = _____

Total watts ÷ system voltage = _____ amperes

11.10.2.2.5. If a shore-power system is to operate on 240 volts, split and balance loads into Leg A and Leg B. If a shore-power system is to operate on 120 volts, use Leg A only.

Leg A / Leg B
_____ / _____ Total Amperes

11.10.2.2.6. Add nameplate amperes for motor and heater loads:

Leg A / Leg B
_____ / _____ exhaust and supply fans
_____ / _____ air conditioners *, **
_____ / _____ electric, gas, or oil heaters*
_____ / _____ 25% of largest motor in above items
_____ / _____ Subtotal

NOTES: *Omit the smaller of these two, except include any motor common to both functions.

**If the system consists of three or more independent units, adjust the total by multiplying by 75% diversity factor.

11.10.2.2.7. Add nameplate amperes at indicated use factor percentage for fixed loads:

Leg A / Leg B
_____ / _____ Disposal—10%
_____ / _____ Water heater—100%
_____ / _____ Wall-mounted oven—75%
_____ / _____ [Stovetop] cooking unit—75%
_____ / _____ Refrigerator—100%
_____ / _____ Freezer—100%
_____ / _____ Ice maker—50%
_____ / _____ Dishwasher—25%
_____ / _____ Washing machine—25%
_____ / _____ Dryer—25%
_____ / _____ Trash compactor—10%
_____ / _____ Air compressor—10%
_____ / _____ Battery charger—100%
_____ / _____ Vacuum system—10%
_____ / _____ [Sum of] other fixed appliances
_____ / _____ Subtotal
_____ / _____ **Adjusted subtotal

NOTE: **If four or more appliances are installed on a leg, adjust the subtotal of that leg by multiplying by 60% diversity factor.

11.10.2.2.8. Determine Total Loads.

Leg A / Leg B
_____ / _____ Lighting, receptacles, and small appliances (from E-11.10.2.2.5)
_____ / _____ Motors and heater loads (from E-11.10.2.2.6)
_____ / _____ Fixed appliances (from E-11.10.2.2.7)
_____ / _____ Freestanding range (see Note 1)
_____ / _____ Calculated total amperes (load)

NOTES: 1. Add amperes for freestanding range as distinguished from separate oven and cooking units. Derive by dividing watts from Table III [opposite] by the supply voltage; e.g., 120 volts or 240 volts.

2. If the total for Legs A and B are unequal, use the larger value to determine the total power required.

TABLE III	Freestanding Range Ratings	
Nameplate Rating (watts)		**Use (watts)**
10,000 or less		80% of rating
10,001–12,500		8,000
12,501–13,500		8,400
13,501–14,500		8,800
14,501–15,500		9,200
15,501–16,500		9,600
16,501–17,500		10,000

Ratings are for freestanding ranges as distinguished from separate oven and cooking units.

add additional service capacity or not.) ABYC Standard E-11 offers a fairly complete section on performing an AC load analysis, and it really mirrors the techniques prescribed by the NEC. The sidebar on pages 69–71 excerpts this material from ABYC E-11.

USING A POWER ANALYZER

I use a WattsUp? portable power analyzer made by Electronic Education Devices. Although the name may sound a little silly, the WattsUp? is actually a high-tech, microprocessor-driven device with great utility. Because wattage or, more precisely, watt-hours are what utility companies base their billing on, the WattsUp? device is quite useful in performing power consumption analysis on many boats. With it, you can:

- Perform power consumption analysis by measuring wattage used.
- Calculate the cost of operation for a piece of equipment. Program in the local utility cost per kilowatt-hour and allow the unit to run through a normal daily service cycle (typically, 24 hours).

- Measure line voltage to an appliance.
- Measure amperage draw, minimum amperage draw, and maximum amperage draw. These values are useful for determining appropriate wire sizes and overcurrent protection device ratings.
- Calculate power factor, which is primarily what I use it for.

The WattsUp? portable power analyzer. Six main modes display the primary data: watts, kilowatt-hours, time, cost, volts, and current.

> **PORTABLE POWER ANALYZER**

WattsUp?, Electronic Education Devices, www.doubleed.com (available through Professional Equipment, www.professionalequipment.com)

Selling for around $150, the WattsUp? is affordable, and costs far less than high-end, scope-type power analyzers, such as the Fluke model 434, which run between $4,000 and $4,500. (Unless your work involves large marina or shipyard power systems and mega-yacht-sized vessels, this cost would be hard to justify, since power analysis is not something that needs to be done all that frequently.) Unfortunately, the WattsUp? works only with 120 V, 60 Hz appliances, so its usefulness is essentially limited to North America.

The unit is simple to operate, requiring only that you plug it in between the appliance in question and any 15 amp receptacle and let it run.

If the appliance is hard-wired into the boat's AC system (which is the case with many battery chargers, for example), simply make up a 15 amp, three-prong plug assembly with an insulated three-terminal strip to temporarily disconnect from the boat and connect to the appliance feed cable. Using this adapter, plug the WattsUp? into a conventional wall socket and attach the three-wire adapter to the appliance via ring terminals at its power input bus and the three-prong plug at the other end of this short jumper (about 12 to 18 inches works fine) plugged into the WattsUp?.

Warning: Turn off all power to the appliance you are connecting this adapter to until all the terminations are made. Shock hazard exists if you inadvertently grab one of the live wires that you've temporarily exposed.

Comparing Power Consumption

I recently used the WattsUp? to compare two 6,000 Btu air conditioners; the newer one was an Energy Star qualified unit, and the other unit was about five years older. I've summarized my results in Table 6-1.

These differences may not seem like much, but on a boat equipped with multiple units cycling all the time, the cumulative difference over time can be considerable.

TABLE 6-1	Power Consumption of Two 6,000 Btu Air Conditioners			
	Average Draw (compressor running; watts)	Peak Draw (at compressor start-up; watts)	Power Factor (fan only)	Power Factor (compressor running)
Older unit	800	1,901	1.0	0.89
Newer, Energy Star unit	560	1,724	1.0	0.98

The power factor rating for the more efficient unit, which is plugged directly into the front of the WattsUp? instrument.

In researching industrial motors like the ones used in marine air-conditioning and refrigeration systems, I discovered power factor specifications ranging anywhere from 0.57 to 0.95—quite a spread in what ultimately amounts to the efficient use of power. But beyond efficiency, power factor also relates to such important issues as the amperage-carrying capability of supply conductors, the susceptibility of equipment to undesirable voltage drop, and the potential for heat buildup and insulation damage due to excessive electrical resistance.

Think of it this way: appliances with low power factor values will use more power (wattage), and typically require more amperage to function. This means that larger feed conductors will be needed to supply the appropriate amperage. Additionally, since more current must be delivered, undersized wiring will contribute to potentially excessive voltage drop. Any AC appliance that is running continually at lower than specified voltage will run hotter than it should, and that excessive heat can ultimately be the cause for breakdown of insulation on motor windings

and the like. The bottom line here is that the longevity of the appliance will be affected and you are going to be wasting power.

A good resource for motor specifications is available through the U.S. Department of Energy at http://www1.eere.energy.gov/industry/bestpractices/software.html. Once you get to the site, you will find a free software download for MotorMaster+ 4.0, which enables you to compare a large number of motors by performance specification and to calculate and compare their cost of use over time.

VALUE AND UTILITY

The WattsUp? is not the first tool the advanced marine electrician or boatowner should purchase, but it does have utility for load analysis of AC systems. It won't work on high-current-draw items like electric ranges and ovens, but for refrigeration systems, microwave ovens, and other commonly used convenience appliances, it can tell you a lot about power consumption characteristics.

Analyzing Harmonic Distortion

AC waveform analysis on an oscilloscope may sound like a pretty sophisticated procedure. And indeed it is, but not for the user, because the sophistication occurs within the instrument itself. It's therefore a practical troubleshooting procedure for any marine electrician or a boatowner committed to running an electrically efficient boat.

When you think of oscilloscopes (commonly called scopes), you may picture the breadbox-size instruments of long ago. But as with all things electronic, the scopes used today have come a long way. Vacuum tubes are ancient history, and transistors and computer chips have taken over, enabling vendors to make some small, yet excellent, handheld devices that really perform. These units are not in the low-priced gear category, although I've seen some advertised for as little as $300. The unit that's shown in this chapter is from the Fluke 860 series, and costs about $2,000. (Note: This has recently been replaced by the 190 series.)

DO YOU NEED A SCOPE?

On a modern boat, AC can be supplied by a variety of sources, and sometimes even a combination of sources. Shore power is only one part of the picture. AC generators, DC-to-AC inverters, and converters that modify incoming voltage and frequency are becoming mainstream—even on boats that historically were considered too small to bother with dockside power.

Unless you do a lot of work on complex AC systems with inverters, generators, and transformers that perform voltage and frequency conversions, an oscilloscope should not be the first tool on your wish list. You can get almost as much utility from a high-end multimeter with a "peak hold" feature and data-logging capability. For those who would like to get into the oscilloscope game without investing several thousand dollars, I recommend the Extech 381 series scopes, which cost approximately $250 to $400.

OSCILLOSCOPES

190 series, Fluke, www.fluke.com (replaces the 860 series)

381 series, Extech Instruments, www.extech.com

printer or PC connection

line or battery power

soft keys

hard keys

rotary selector

test leads

The Fluke 867B graphical multimeter is a full-feature multimeter and oscillo-scope. It also has the ability to log data over time, and upload it to a PC for further analysis, something we'll address in Chapter 8.

Unfortunately, some of the electronic equipment routinely installed on boats can deliver less than satisfactory performance when powered by distorted AC sine waves. For example:

- flat-screen televisions—may display diagonal lines
- some audio systems—may generate annoying background noise
- regular TV screens—may develop an audible "hum" and display diagonal lines
- some computer systems, especially systems using a UPS (uninterruptible power supply)—may work but run too hot; may try to turn on and off; or may not work at all (note that all the laptops I've tried work fine with an inverter; some desktops may be affected)

A classic symptom of a faulty waveform is a piece of gear that functions perfectly when run from the shore-power supply, but acts up when run from an inverter or generator.

The good news is that with an oscilloscope you can see and analyze the AC waveform to detect harmonic distortion (see below).

AC LOADS

Let's look again at the part of the power factor definition in Chapter 6 that stated "a unity power factor (1.0) indicates that the current is in phase with the voltage." To understand this, we need to look at some additional definitions and take a closer look at what happens in an AC circuit depending upon AC load characteristics.

ALTERNATING CURRENT AND WAVEFORMS

AC current forms *sine waves* when voltage is charted against time. The waveform displayed on the oscilloscope is a *true sine wave*. Although the nominal voltage is 120 volts at 60 cycles per second, the scope shows the voltage as slightly less than that, at just over 117 VAC. This is completely normal; voltage variations of ±10% are not uncommon.

Some of the equipment we use on boats doesn't always deliver a perfect sine wave, however. Generators sometimes develop mechanical or electrical problems that can affect their ability to maintain precise engine rpm, and frequency and voltage, as well as the waveform, can become skewed. The oscilloscope will display this faulty waveform, as well as frequency and voltage values that fall out of acceptable parameters (generally ±10% for voltage and ±5% maximum for frequency). Many inverters don't even attempt to deliver a perfect AC sine wave, but rather will deliver AC current in a *stepped-square-wave*, or *modified-square-wave*, output. Charles Industries, one major vendor of inverters, offers at least one unit that bills itself as a quasi-sine-wave inverter. This appears to be more marketing hype than anything else. The bottom photo shows the waveform as delivered by one of these inverters. These inverters show these "stepped" waveforms because the unit is taking DC current, which shows as a flat straight line on an oscilloscope, and converting it to what I'll call a usable form of AC current. The "steps" are created by the electronic switching that is occurring in the inverter, which is really nothing more than the power turning on and off as it gets "stepped up" to AC voltage potentials. More sophisticated inverters will have more electronic switches, which in effect creates less profound "steps"

A perfectly formed AC sine wave, characteristic of AC power as delivered by a utility company.

A modified square wave, or "quasi-sine" wave, as Charles Industries calls it, the company that distributes the DC-to-AC inverter this reading was taken from. These quasi-sine waves represent a certain level of distortion. If you have an inverter, you have to figure out whether the appliances you're attempting to power with the inverter can operate in spite of the inverter's distortion.

A comparison of inverter waveforms. (Reprinted with permission from Boatowner's Illustrated Electrical Handbook, second edition, by Charlie Wing)

and will smooth out the curve, or wave. It is important for boatowners and technicians to be aware of this because some equipment requires true-sine-wave output to function properly, as discussed on page 75.

Load Types

AC loads can be divided into two categories: linear and nonlinear. *Linear loads* basically draw current in proportion to the voltage delivered, much like a DC load where Ohm's law applies. That is:

$$\text{amperage} = \text{volts} \div \text{ohms}$$

A basic incandescent lightbulb is a good example of a linear load. There is a voltage differential across the bulb's filament, the filament has an engineered amount of resistance, and the resistance generates enough heat to make it glow brightly. The relationship between the voltage delivered and the resistance of the filament determines the current draw through the circuit.

Nonlinear loads, which work on the principle of inductance, include devices such as transformers, motors, solenoids, and relays. These have a more complex relationship between volts and amps.

First, a magnetic field develops around any wire or conductor through which current is flowing. The strength of the magnetic field is proportional to the amount of current (i.e., amperage).

Second, we also know that we can induce electrical current flow through a conductor by moving a magnet rapidly in close proximity to the conductor. A transformer is simply a pair of wire coils in close proximity to one another, with an iron core in the middle to contain the magnetism. As we pass alternating current through one of the transformer's coils (the primary winding), an alternating magnetic field is created. The field rapidly rotates around the secondary winding, and electrical current flow is induced (at 60 times per second, or 60 Hz, in North America, and 50 Hz in most of the rest of the world). Sounds perfect, right? Well, it isn't quite.

In fact there is a loss through the transformer, resulting in some inefficiency. The level of this inefficiency is a function of harmonic voltage levels at a transformer's primary winding versus load-generated harmonic current (nonlinear) levels at the output terminals on the secondary side of the transformer, and a phase relationship is established. To explain more simply, remember that both voltage and amperage in AC circuitry are not constant as in DC circuitry. Alternating current (and voltage) is delivered as waves—there are peaks and valleys. When we measure 120 VAC with a multimeter, the actual peak voltage is usually on the order of 165 to 170 VAC. But it does vary slightly due to forces within the devices that are powering up, which either slow things down a bit or speed them up. When this happens, the waveforms become slightly distorted, and we end up with slightly skewed waveforms. The term used to describe this effect is *harmonic distortion.*

HARMONIC DISTORTION

Harmonic distortion is defined in the National Electrical Code as "a load where the wave shape of the steady-state current does not follow the wave shape of the applied voltage." As we just discussed, the bottom line is that nonlinear loads—such as audio equipment, computers, fluorescent lighting with ballast, and variable-speed motors—will put AC-supplied voltage out of phase with amperage. The net result is a waveform that becomes slightly distorted when viewed with an oscilloscope. Linear loads—such as a water heater or an incandescent lightbulb—will not create this phase imbalance between voltage and current. But all of this is quite esoteric, and really drags us into advanced electrical engineering concepts that go beyond what I am trying to convey here. What's important for the boatowner or technician to understand is that excessive harmonic distortion can create problems with some common onboard AC equipment.

The illustrations show a change in the relationship of voltage and current in an inductive load scenario (top illustration), which could be a transformer or a motor, and in a capacitive scenario (bottom illustration), which could be a noise filtration circuit on something like an audio amplifier. In the inductive scenario, the voltage is leading the current. In the capacitive scenario, the current leads the voltage. These are typical scenarios for these broad load categories, but don't get too caught up in all of this. Again, what the boatowner or technician needs to know is that things are

Voltage (V) is leading the current (shown as amperage, I) because it takes time for the voltage to force the buildup of current to its maximum across an inductive load.

Another example is what happens in a capacitor as it charges and discharges. In this case the voltage lags behind the current because the current must flow to build up the charge in the capacitor. The effect is known as capacitance, which is the ratio of the electric charge transferred from one to the other of a pair of conductors to the resulting potential difference between them.

ending up a bit out of phase and waveform distortion will occur as a result. (Remember that resistive loads do not have these faults and are generally thought of as linear; there is no inductance or capacitance to worry about.)

Regardless of whether an AC circuit is supplying an inductive load, a capacitive load, or a resistive load, there are factors that will modify the perfect AC current sine waveform.

Excessive harmonic distortion may cause equipment to overheat, motor failure, capacitor failure (especially in motor circuits), and excessively high, neutral conductor current. Commonly used devices that may either contribute to or be affected by excessive distortion include:

- fluorescent lights
- some HVAC systems
- computers, printers, and fax machines
- dimmer switches for lighting
- audio equipment
- video equipment
- UPS devices

Here's an example of how harmonic distortion is created. In order to work, a fluorescent light first requires a buildup of voltage. Once the voltage is high enough, an arc occurs across its pair of electrodes. At this point, current flows easily, especially when compared to the current flow before the arc occurred. In effect, the light draws nearly no current initially and much more current once the tube is activated and producing light.

Multiply this effect by a power grid's cumulative nonlinear loads, and the net effect is a distorted sine wave delivered to the boat. Extreme distortion can be seen on an oscilloscope visually as a sine wave that has uneven peaks and valleys, often with a "flattened" top to one of the wave peaks (see top photo page 81).

Harmonic distortion can be cumulative within a power grid, and will be distributed throughout the grid, potentially causing equipment problems. Determining whether

excessive distortion exists is one of the more advanced troubleshooting exercises a boatowner or technician may find themselves performing at some point.

Finding Harmonic Distortion

Finding the source of harmonic distortion is not always easy. Harmonics can migrate through the power distribution system from one building or marina to another, and from one boat to another in a marina, if they are connected electrically via the utility power grid. Furthermore, the problem can be intermittent, as electrical loads that cause distortions may cycle on and off. Monitoring must, therefore, be done over an extended period of time. In fact, some power analysis specialists recommend site monitoring over a period of one month to gather quantifiable data! This makes a strong case for the data-logging methodologies we'll look at in Chapter 8.

If you have equipment that is exhibiting harmonic distortion symptoms—the most common ones are overheated or damaged AC neutral conductors or cable terminations, and circuit breakers that trip continually and mysteriously—you can use your oscilloscope to confirm your suspicions.

Connecting an oscilloscope to an AC circuit is easy. Plug the meter's red lead into the meter's AC volts socket and the black lead into the "com" socket, as in the photo on page 75. Then connect the other ends of the leads as if you were measuring voltage: red lead to the hot terminal, and black lead to the neutral terminal. Next you'll select the appropriate function for the meter you are using.

Calculating Distortion Using Peak Capture

Although special power-analyzing meters are currently available, they are cost prohibitive, especially when you consider that even advanced marine electricians will not be using this equipment that frequently. But with an oscilloscope, or with a high-end DVOM that also has a "peak capture" feature, you can get a pretty good indication of whether excessive distortion exists in an AC power supply. This "peak" is the maximum point we see when looking at the sine wave on an oscilloscope; i.e., the highest voltage measured in the AC waveform.

Crest factor describes the ratio of the peak value of a measured waveform to its root mean square (RMS). A pure, undistorted sine wave's crest factor, therefore, is described by the equation:

$$1 \div 0.707 = 1.414$$

In other words, the peak voltage value of an undistorted sine wave is 1.414 × the true RMS value. Any variation from this value indicates a distorted waveform and bad harmonics. Using a peak-capture-capable DVOM, here's how to perform these measurements, step by step:

1 First, measure the true RMS value of the voltage at a receptacle. When using a true RMS meter, the reading will be automatic if you are set to the VAC scale on the meter.

2 Multiply that reading by 1.414 to establish the theoretical peak value.

3 Check the actual peak value with the peak capture feature on your meter.

4 Compare the actual value with the theoretical value. If they differ by more than 2% or 3%, excessive harmonic distortion exists.

To differentiate between voltage and current harmonics, consider these points:

• For voltage harmonics, the typical crest value will be less than the product of the RMS value × 1.414. On an oscilloscope, this will appear as a "flat top" on a sine-wave curve, as shown in the top photo opposite.

• For current harmonics, the typical crest value will be greater than the product of the RMS value × 1.414. On an

oscilloscope, this will appear as an abnormally high peak.

As an example, let's use the numbers displayed in the bottom left photo opposite to calculate distortion. To keep things simple, we'll round the RMS value down to 121.

We begin by determining the theoretical peak by multiplying the RMS value by 1.414:

$$121 \times 1.414 = 171 \text{ VAC}$$

Our actual measured peak (see Peak Max on scope), at 161 V, is lower than that value, which would appear as a flat-top waveform on an oscilloscope.

Subtract the measured peak from the theoretical peak:

$$171 - 161 = 10 \text{ V}$$

Divide the difference by the theoretical peak:

$$10 \div 173 = 6\%$$

Since 2% to 3% is considered the allowable limit, this system has excessive distortion.

Current Multiplication

Excessive current in neutral conductors is a common phenomenon in the three-phase power supplies (where three sinusoidal voltages are generated out of phase with one another) typical of many marinas. Before we leave this discussion of harmonic distortion, we need to understand how and why it occurs.

In a well-balanced, three-phase AC circuit (which is how all AC power begins in shore-power installations), there are always harmonics as the wave of the current alternates. Even-numbered harmonic currents cancel each other out; odd-numbered harmonics do not—in fact, they add algebraically, causing the neutral conductors in a three-phase system (the white wire in North America) to carry as much as 180% of the amperage carried by each of the three individual phase conductors. This particular problem has been identified in

A typical distorted flat-top waveform.

A peak value result on an oscilloscope.

A typical true RMS voltage reading for a 120 V system.

Delta configuration—primary side of transformer

Wye configuration—secondary side of transformer

B

A

C

"WYE"

B

A

neutral

C

208 V

208 V

208 V

120 V

120 V

120 V

A Delta-to-Wye transformer showing the differences in configuration between the primary and secondary windings of the unit.

office buildings for some time, and may account for the seemingly strange symptom of a burned-up neutral conductor with all other conductors in the system looking fine.

The most common transformer type used in commercial and industrial environments is known as a Delta-to-Wye transformer, in which the transformer's primary windings are in a Delta configuration, and the secondary windings are in a Wye (see illustration). We can see that the neutral has to serve each of the three hot-wire legs to get the current back to its source; i.e., the transformer. Since this neutral conductor is typically the same gauge as the hot wires, it's easy to see how it might be overburdened at times.

Transformer overheating is not only another effect of harmonic distortion exposure, it presents another potential problem on boats that use transformers for shore-power isolation and/or as voltage step-up or step-down devices. The infrared heat gun discussed in Chapter 5 is a great tool for determining whether a problem exists. Temperature readings greater than 120°F to 125°F (49°C to 51.7°C) indicate either a transformer that is underrated for the task at hand or is exposed to excessive harmonics.

Harmonic distortion was of little concern to boatowners and technicians when the AC load on boats was typically only a water heater, perhaps an electric stove, and maybe a TV. But as we've discussed, the equipment on boats has gotten more sophisticated, and consequently more sensitive to distortion in its power supply. So it is now important for the advanced technician or boatowner to have an understanding of harmonics to ensure the safety and reliability of AC systems.

VALUE AND UTILITY

If you regularly work on larger boats or boats with AC generators, inverters, or any combination thereof, I recommend that you consider investing in an oscilloscope. If time is money, and most say it certainly is, then you'll find that a small scope will answer some key questions about power supplied to the boat *very* quickly and easily. As boats get more and more sophisticated, and we find ourselves working on 35-footers equipped with multiple AC power sources, flat-screen TVs that rise up out of a saloon table, and high-end audio systems, etc., the value of using a scope will become evident in no time at all when problems occur.

Using a Laptop for Intermittent Problems

One of the first troubleshooting steps is confirming or denying the existence of a problem, and one of the most difficult troubleshooting challenges is the intermittent problem. You know—the problem that doesn't show up when you want it to. The troubleshooting process also requires that you understand the conditions under which the problem occurs. Unfortunately, at today's labor rates it is totally impractical to achieve these goals in many cases. Who wants to pay a technician hundreds of dollars just to sit around waiting for a problem to occur? And few boatowners have the time to conduct extensive in-use testing with the sole objective of waiting for an intermittent problem to recur.

One way to deal with this situation is to let your equipment do the sitting around for you, gathering data that you can analyze later. Sometimes environmental conditions trigger intermittent electrical problems and often these conditions can be duplicated as part of a sea trial. Simply attach data-logging equipment to the circuit(s) in question and put the boat through a sea trial, running the boat's systems through their normal operational cycles while the data-logging equipment does its work. Keep a log during the sea trial, noting sea conditions, wind velocity, engine rpm, boat speed, precipitation, and even temperatures in the area of the meter at given times throughout the test. This will enable you to later match the environmental conditions to the data-log time stamp on your laptop as an aid to analysis of the data. After the sea trial, check the measurements and key them to events during the sea trial.

LAPTOP LINK-UPS

Like so much of the equipment discussed in this book, the microchip is the driving force behind the data-logging capabilities of just about any piece of high-end electrical measuring equipment you are likely to buy today. The software involved is not complex, so it doesn't need the latest and greatest in a PC or laptop. I use an otherwise outmoded PC as my platform for data logging.

Typically, a PC that is at least Windows 95 compatible will suffice. You won't need much storage either, as all the logging programs I'm familiar with are essentially text based and don't require a huge hard drive or lots of memory.

A somewhat outmoded laptop or PC capable of running Windows 95 or better is generally sufficient to capture output from most equipment into data-logging software.

Multimeters and other instruments with the capability to record measurements to internal memory or upload them to a computer are available from equipment suppliers such as Fluke, AEMC, and Extech. You will need the appropriate software from the meter vendor. This may be included in the price of the meter or available as a low-cost option (typically in the range of $30 to $60). Additionally, you will need a data-link cable to connect the meter to your PC, which is generally included with the software and in most cases is a serial port connection. By following normal PC software installation procedures, and selecting a communications port for your PC, you should be ready to go in no time.

Most meters can be programmed to record different intervals of data "hits." For example, my Fluke 860 with FlukeView software allows me to program data sampling in 1- to 45-second intervals or 1- to 15-minute intervals. Depending upon what you are checking—rapid change or change over extended periods of time—this is an important adjustment. Also, if you are monitoring over a 24-hour period, you may want to extend the sample-rate time period to conserve hard drive space on your PC.

No adjustments are required in how you connect the meter to the circuit—volts, amps, and ohms are tested the same way as with non-data-logging instruments. The only essential difference is the use of spring-loaded alligator clips instead of straight test probes to hold the leads in place while the meter goes to work.

Additional Advantages

While our primary focus is identifying intermittent problems, having a PC or laptop linkup has other advantages. For example, the larger screen of a computer versus a meter magnifies your view dramatically. This is particularly useful when looking for distortion in AC waveforms. Even if your multimeter oscilloscope has the technical capability to capture subtle problems, the size of the display (2 to 4 inches, typically) may make it difficult to observe. But if you blow up the view on a 14- or 15-inch PC or laptop screen, those problems become obvious. For example, the flattening at the peak of the wave shown in the top photo on page 81 is easy to see when magnified on a computer screen but would be hard to see on a 2- to 4-inch multimeter display.

If you are a professional technician, sending data logs to a computer opens up several possibilities. You can save and print customer files, include screen shots in a customer's report, or improve your internal documentation and quality control processes. It's easy to do, and it adds a level of professionalism to your work that is unprecedented in the marine field.

REAL-WORLD EXAMPLES

Let's look at some real-world cases where you might consider data logging as a useful aid in troubleshooting.

Example 1: Unexplained Underwater Metal Corrosion

One of the classic problems affecting boats in marinas is rapid, unexplainable corrosion of underwater metal components. Often, when going through a complete corrosion analysis, you may find no measurable problem. The problem could be cycling of appliances or equipment, such as an air conditioner or bilge pump, either on your own boat or one nearby at the dock. So what do you do?

You want to capture stray current in the water surrounding the affected boat "after hours." Link a tracking multimeter, fitted with a special silver/silver chloride reference electrode (see Chapter 11), to your computer, and immerse the electrode in the water near the affected boat. The computer link-up will record the data, allowing you to either confirm or rule out the presence of stray current. If there is stray current, you'll be able to pinpoint the exact time and date of the measurement. More important, you will have confirmed a probable cause of corrosion, without having to make five repeat trips to the boat to try and "catch" the problem. And, if you're a technician, you avoid having to report to the client that "no problem was found at the time of survey." Or if you're a do-it-yourselfer, you won't have to hang out on the boat all night, looking at the multimeter readout every 10 minutes.

Example 2: Instrument Lighting That Fails Only When Underway

In this example, the backlighting on your dash-mounted engine monitoring instruments blinks on and off when motoring in rough sea conditions. Most people would immediately attribute this problem to a loose electrical connection feeding the lights. But a look at the electrical feeds to the dash lights has come up with nothing, and the size and complexity of the boat's overall wiring make it difficult to inspect every connection for every wire. Further complicating the diagnosis is the fact that the instruments are backlit by LEDs mounted on a printed circuit board. The problem might, therefore, be in the power supply or ground return for the PC board, or it might be on the board itself, which is sealed and integrated with the backside of the instrument housing.

To troubleshoot this problem, you should log data while reproducing the conditions in which the problem occurs. Attach a data-logging multimeter to the power supply and ground it to the instrument cluster. Then take the boat for a sea trial in rough seas until the dash lights begin to blink. You can then confirm whether or not the power feed to the printed circuit board is affected. If not, and the problem occurs, it's a safe bet the problem is in the PC board itself, which in all probability needs to be replaced.

By using this methodology you have turned a two-person job (one to operate the boat, the other to watch the meter face) into a one-person/one-instrument job and created a situation where you can run the test without any special scheduling or logistical arrangements. Plus you have isolated the location of the problem without employing frustrating trial-and-error tactics.

These methods are truly a more efficient approach to problem solving. Boatowners doing their own work will find answers more readily, and for the marine professional, this efficiency will pay off in less wasted time and better customer service.

A sample of a DC log. Notice the time and date stamps with each entry. All you need to do is scan these data entries and pick out the odd entries, such as those marked, then try to associate the time they occurred with events that might have caused them. For example, if a problem occurs on the unoccupied boat between 5 P.M. and 8 P.M., it could be associated with the use of electric galley stoves on other boats at the dock. If it occurs at fairly regular intervals, but only during hot weather, it might be an air-conditioning unit.

Testing for Electromagnetic and Radio Frequency Interference

As boats become more complex, and we continue to install more and more electrical and electronic gear on board, the potential for electromagnetic and radio frequency interference (EMI and RFI) increases. When you combine this with the space restrictions on boats, you'll see the challenges installers face in finding suitable locations for all the equipment carried by the average boat today.

UNDERSTANDING ELECTROMAGNETIC INTERFERENCE

The ABYC electrical standards state that wire and cable runs need to be thought out when they are being installed in close proximity to "magnetically sensitive devices," such as the boat's compass. But the truth is that we need to consider far more than just the compass and magnetic sensitivity. For example, all microprocessors use clock oscillators that create a digital square waveform (see Chapter 7). This waveform is not only loaded with distortion and harmonics, it is affected by stray emissions as well. Devices such as fluxgate electronic sensors (used as direction sensors on autopilots), satellite TVs, and satellite telephones are also extremely sensitive to excessive interference. Tales abound of electronically controlled engines displaying strange running conditions—such as running rough or excessive black exhaust smoke—when certain, seemingly unrelated, electrical devices were activated. Certainly over the years there have been cases where electronic navigation systems have been brought to their knees due to RFI (Loran-C was particularly vulnerable), and I've even heard stories of lights mysteriously turning on and off when someone keyed a single-sideband radio to transmit.

These strange symptoms are more easily understood in the light of two basic concepts:

1 All wire and cable that have electricity moving through them will have a magnetic field surrounding them.

2 The strength of this magnetic field is directly proportional to the amount of electrical current (amperage) flowing through the wire or cable; i.e., more amps = a stronger magnetic field.

In the field of electronics, EMI and RFI are synonymous—they both arise from electromagnetic radiation, which is energy made up of electromagnetic waves. These waves are classified by their frequencies, aligned on a spectrum ranging from low to high: radio waves, microwaves, infrared radiation, visible light, ultraviolet radiation, X-rays, and gamma rays. The nature of electrical current is to emit electromagnetic radiation. This energy becomes interference when it causes unwanted signals or noise to occur in other devices or circuits that are operating in or close to its frequency range.

FCC-REQUIRED VERBIAGE FOR CLASS B DEVICES

This equipment has been tested and found to comply with the limits for a Class B digital device, pursuant to part 15 of the FCC Rules. These limits are designed to provide reasonable protection against harmful interference in a residential installation. This equipment generates, uses, and can radiate radio frequency energy and, if not installed and used in accordance with the instructions, may cause harmful interference to radio communications. However, there is no guarantee that interference will not occur in a particular installation. If this equipment does cause harmful interference to radio or television reception, which can be determined by turning the equipment off and on, the user is encouraged to try to correct the interference by one or more of the following measures:

- Reorient or relocate the receiving antenna.
- Increase the separation between the equipment and receiver.
- Connect the equipment into an outlet on a circuit different from that to which the receiver is connected.
- Consult the dealer or an experienced radio/TV technician for help.

STANDARDS AND REGULATIONS

Most countries require that electronic and electrical hardware work correctly even when subjected to specified amounts of RFI, and that they should not emit RFI that could interfere with other equipment. The U.S. Federal Communications Commission (FCC) has several applicable regulations addressing these emissions. Manufacturers must test and label digital devices to inform end users of the maximum emission level from a product when it's used in accordance with its instructions.

The FCC has established two classes of devices. Class A equipment is intended for use in industrial or commercial areas. Class B, which is more stringent, includes residential applications as well as industrial applications. As shown in the sidebar, the regulations include a disclaimer that must be stated in the instruction manual, essentially stating that harmful interference, while prohibited, may occur anyway.

The electronics installation standards of the National Marine Electronics Association (NMEA) mirrors the FCC's advice.

What the equipment manufacturers aren't telling you is that the FCC requirements were intended to deal with issues related to industrial and residential interference migrating from one business or household to another, and not for the relatively close proximity of gear installed on boats! So even though these standards exist, we still run into problems on boats due to RFI noise and/or strong magnetic fields.

COMMON SOURCES AND VICTIMS OF INTERFERENCE

Following are some of the most common sources of EMI/RFI emissions on boats:

- AC generators
- air conditioners

- battery chargers
- blower fans
- DC-to-AC inverters
- depth sounders
- electric fuel pumps
- engine alternators
- engine ignition systems
- engine starter motors
- engine tachometers
- fans
- fluorescent lights
- high-amperage wire runs (e.g., power feeds to bow thrusters, anchor windlasses, and electric winches)
- lighting transformers
- radars

- shore-power isolation transformers
- trim tab motors

That's a fairly long list of culprits. And what are the common victims of EMI/RFI?

- AM/FM stereos
- audio systems
- autopilot direction sensors (fluxgate sensors)
- computerized engine controls and their wiring harnesses
- flat-screen TV monitors
- depth sounder transducer cabling
- DGPS receivers
- Loran-C receivers
- magnetic compasses

interference conducted to receiver

interference re-radiated to aerial and D/F loop

interference radiated to ship rails

interference radiated from loose connections

interference conducted through ship's wiring

interference from ignition

interference generated by propeller shaft

Interference is caused by unwanted radio frequency energy. The energy is picked up by antenna systems and wiring circuits and fed to receivers, which may not be able to distinguish between noise and signals if they share similar frequencies. (Reprinted with permission from Boatowner's Mechanical and Electrical Manual, third edition, by Nigel Calder)

- microprocessors
- SSB radios
- TV and satellite tracking antenna direction sensors
- VHF radios

So when equipment misbehaves, how do we find the source of interference? I have two tools I like to use.

TOOLS OF THE TRADE

Even though EMI and RFI are considered synonymous, for the purposes of troubleshooting, I divide them into two categories: RFI, which I can hear; and EMI, which I can't hear, but I can measure. I use two tools—one somewhat sophisticated, and the other extremely unsophisticated—to search out the potential for EMI/RFI emissions to create problems. To measure EMI, I use a gauss meter. To listen to RFI, I use a simple transistor radio. Using both of these two tools is extremely easy.

Transistor Radio

Let's start with the transistor radio. Pull out the antenna if it has one and turn up the volume as high as you can. Set it to the AM band and tune it to just off the scale, either high or low, so that no sound can be heard other than perhaps some low-level static. That's it—now you are ready to try it out. Here's how:

1 Turn on a battery charger or use the alternator on your car.

2 Gradually bring the radio closer to the charger or alternator; you will clearly hear static getting louder and louder.

3 Move the radio back and forth and all around the interference source to be sure you don't miss any leakage noise.

4 After you get the noise to peak, gradually move the radio away from the source. The distance at which you no longer hear the interference becomes the minimum safe distance (the separation distance or zone) from the unit that you should mount sensitive equipment.

5 Repeat this procedure with any interference source on the boat.

I use an AM/FM transistor radio that I picked up for $8 at Wal-Mart to listen to RFI emissions in both the kHz range and part of the MHz spectrum.

To establish a safe RFI separation distance or zone from an alternator, move the radio around until the radio static peaks. Then gradually move the radio away from that point until you no longer hear static. This is the minimum separation distance.

TABLE 9-1	Operating Frequencies of Common Onboard Electronics	
Electronic Equipment		**Frequency**
Cell phone		850 MHz or 1.96 GHz
Depth sounder		50 or 200 kHz
Differential GPS		150–500 kHZ
GPS		1.5 GHz
Loran-C		90–110 kHz
AM radio receiver		540–1,700 kHz
FM radio receiver		88–108 MHz
SSB radio		2–22 MHz
VHF radio		160 MHz

As we saw above, frequency matters, and on your transistor radio, the AM band you are listening to is in the 540 kHz to 1,700 kHz range. If you switch to the FM band and use the same technique, you will be listening in the 88 MHz to 108 MHz range.

You'll only encounter interference problems when the interfering frequency is fairly close to the frequency at which a given device operates. What frequencies do boat appliances run at? Table 9-1 lists the operating frequencies of some common marine electronic devices. The frequency range of the radio is close enough to the frequencies of most onboard electronics to reveal interference problems.

Once you've figured out where the interference is coming from, turn to Chapter 12 to find out how to minimize its effects.

Gauss Meter

The gauss meter is used in almost exactly the same way as the transistor radio, except that instead of listening to the interference, you get a numerical readout in milligauss (mG). For the meter to measure the magnetic field strength accurately, the device being measured must be pulling full current. As you move the meter closer to a device, the readings will

This Sperry gauss meter has a range of 0.1 to 199.9 mG.

GAUSS METERS OR ELECTROMAGNETIC FIELD TESTERS

EMF-200A, Sperry Instruments, www.awsperry.com
Model 480823, Extech Instruments, www.extech.com

increase if interference is present, and decrease as you move away. As with the radio, move the meter around, trying different angles, to determine the peak of the interference, then move it away from the device to find the minimum separation zone.

I use a gauss meter to measure relative magnetic field strength at various points around a conductor or piece of equipment, as well as to probe near large DC conductors and equipment that delivers high amperage to high-current-draw motors, such as starters and bow thrusters. It is a handy tool, costing about $100.

Using a gauss meter to check the separation zone near a high-current-draw cable from a battery switch. Notice the reading of 2.5 mG. The circuit in question (a bilge blower fan) was drawing 6 amps, which is not a particularly high current draw, but I wouldn't mount a digital fluxgate compass sensor near this wire.

Here I used a small handheld compass to see how much compass deviation I could induce, and compared that to the mG reading shown in the previous photo. A mere 2.5 mG produced a 12-degree error on the compass. It took 10 inches of separation before the compass returned to the correct reading. By comparison, some battery chargers and inverters I've checked emit as much as 170 mG! That's enough to send a compass into an extreme level of deviation from the actual heading. At a 170 mG reading, a typical separation zone would equate to 3 to 6.5 feet (1 to 2 m).

Testing for Residual Current and Isolating Its Sources

In a perfect world, alternating current does just that: it alternates back and forth, and it does so evenly, between the hot and neutral conductors (or in the case of 240 VAC, between the L1 and L2 conductors). In effect, the current coming down the hot wire should be exactly equal to the current flowing back down the neutral conductor. The currents cancel each other out because they are perfectly balanced.

On many AC appliances, however, there is always a small amount of current leakage, also called residual current, which finds its way into the grounding conductor. For example:

- appliances with motors, such as refrigerators and air-conditioning units
- appliances equipped with line noise filters, such as battery chargers and stereo amplifiers
- any device that uses power factor correction circuitry, such as, again, refrigerators and air-conditioning units (see Chapter 5)

Additionally, as appliances age, materials such as heating elements and the winding insulation in motors and generators gradually degrade. These conditions also cause residual current.

In this chapter we will discuss both sources of AC leakage and how to test for it using two instruments: a clamp-on AC leak tester and a megohmmeter.

AC LEAKAGE FROM APPLIANCES

Residual current from appliances is a normal situation and not dangerous as long as the level of leakage is maintained at an extremely low level. The International Electrotechnical Commission (IEC) limits acceptable leakage to 0.01 mA to 0.75 mA, depending on the specific appliance. In the United States, Underwriters Laboratories specifications emulate the IEC standard. Suffice it to say that acceptable leakage is limited to extremely low levels.

However, with multiple appliances on board, the normal leakage from each appliance adds up, and this cumulative leakage ends up traveling along the grounding conductor. Low-level AC current leakage begins to occur, allowing yet more AC current into the grounding system. This is where the danger begins. This leakage is often well below the

TABLE 10-1   Effects of Electrical Shocks

Effect	Direct Current (mA)		Alternating Current (mA)	
	Male	Female	Male	Female
Slight sensation	1	0.6	0.4	0.3
Perception threshold	5.2	3.5	1.1	0.7
Shock, not painful, no loss of muscular control	9	6	1.8	1.2
Shock, painful, still no loss of muscular control	62	41	9	6
Shock, painful, "let go" threshold	76	51	16	10.5
Shock, painful and severe, uncontrolled muscular contraction, severe breathing difficulty	90	60	23	15
Shock, possible ventricular defibrillation following 3 seconds of exposure	500	500	100	100

WHY IS RESIDUAL CURRENT LEAKAGE SO DANGEROUS?

Residual current leakage into the water surrounding boats has risen to the top of the safety hazard list in recent years because we now have indisputable proof that it has caused deaths in freshwater lakes and rivers. The concern is not so great in salt water, because salt water is a good conductor of electricity, and any leakage current is rapidly dissipated to earth, either on the dock or the bottom.

Fresh water, however, is a rather poor conductor of electricity. As current leaks from the bottom of a boat, a voltage gradient tends to stratify near the surface of the water, migrating sideways around the boat and seeking a path to earth on the shore. Because the human body is composed of a very high percentage of salt water, a swimmer who enters this voltage gradient and aligns his or her body with the gradient becomes a far more efficient current path than the water itself. Once exposed to this current, a fatal heart muscle seizure may occur, and the swimmer is electrocuted.

How much residual current is potentially lethal? I use 30 mA as my baseline, simply because that is the established standard everywhere in the world except North America. The 30 mA figure is less than the AC figures given for ventricular defibrillation in Table 10-1, but it has been established almost globally as a reasonable compromise between protection devices tripping constantly and extreme pain for humans. Anything above that limit is, as far as I'm concerned, simply too great a risk to take.

level it takes to trip a conventional circuit breaker, nominally rated at 15 or 30 amps, but more than enough to be lethal.

How much is lethal? Table 10-1 describes the typical effects of exposure to DC and AC current on healthy humans (individual responses will vary). Note that these figures are not in amps, but in milliamps!

USING A CLAMP-ON AC LEAK TESTER

To protect against this potentially fatal problem, we need to accurately measure very low levels of AC current leakage, in resolutions of milliamps. The amp clamp provided with the typical multimeter does not have the required level of resolution; typically they can only read down to a minimum of about 0.1 amp. A special tool is required, namely, a clamp-on AC leak tester, which can read down to 0.001 amp.

I use a Yokogawa clamp-on leak tester, model 300-31, for which I paid just over $400. Similar devices are available from other suppliers, but I like this one because its clamp has a 40 mm inside diameter, which will fit around the largest shore-power cords.

Before you begin using a milliamp AC leak tester, it is essential to understand that you will now be measuring amperage in a way that may seem dead wrong. Years ago, we were taught to measure amperage by connecting the meter in series with the load, sometimes using a shunt. We also learned that we could check only one wire at a time. Later, as inductive amp clamps

The Yokogawa model 300-31 AC leak tester.

became popular, we learned that you could only measure amperage by clamping one wire at a time.

Now, however, you will initially clamp your meter around a boat's shore-power cord to check residual current. What's being measured here? It's the differential between what's going down the AC hot wire and what's coming back up the neutral conductor. This differential is current leaking into the grounding conductor, measured in amps. If the reading is zero amps, then no leakage exists, and the hot and neutral are balanced perfectly, as they should be in a perfect situation. But if there is measurable current, the question becomes, where is this

CLAMP-ON LEAK TESTERS

Model 300-31, Yokogawa, www.yokogawa.com
Model 380942, Extech Instruments, www.extech.com

Use a milliamp AC leak tester to check residual current by clamping it around the shore-power cord. This reading of 6.62 mA is fairly low, but still more than desired. I prefer to see readings of 3.0 mA or less.

current going, or, for that matter, where is it coming from? That's the tricky part, because the answer to the question is, it depends . . .

Electrical theory tells the electrician that current will always follow the path of least resistance, which is essentially true. Therein lies the problem. With a boat plugged into shore power, there are multiple paths for residual current to travel. In a perfect world, residual current will travel down the grounding conductor and back to the power source somewhere at the head of the dock. Unfortunately, it is not a perfect world, and dock wiring problems are rampant. We cannot rely on the integrity of the dock's grounding system (which is why I use a SureTest device to help confirm or deny this integrity—see Chapter 4).

Tracing the Source of Residual Current

Tracing the exact source of residual current is, unfortunately, a complex process. It requires process of elimination, which can be exacting, tedious, and at times extremely inconvenient.

I say inconvenient because the possibility exists that another boat may be the source of the current, which is using the superior grounding (compared to the dock) of your boat as the quickest path to the water. If that is the case, then some of that current is surely leaking out through the bottom of your boat.

The potential for current to leak into the water is, ironically, related to ABYC standards, which call for the boat's AC and DC grounding systems to be linked at the boat's grounding bus or at the engine negative terminal. If the AC grounding system on the dock has excessive resistance, some or all of the residual current will exit the boat right through the bonded through-hull fittings as it seeks a path to earth. In other countries, this bonding of the AC and DC grounding systems is not a common practice, nor is it required under international standards. The reason is that they employ the 30 mA whole-boat protection system, which will simply shut down the boat if the residual current hits 30 mA.

To locate the source, do the following:

1 Shut off the main AC breaker to the boat. If the residual current reading goes to zero, then it's a fair bet the source of the current leakage is on board the boat.

2 So next, turn the main breaker back on.

3 Then turn off the breakers to the boat's various branch circuits one breaker at a time.

4 If the current reading suddenly goes to zero when you turn off a breaker, you've identified the problem circuit. Most often, the leakage current is coming from an appliance on the circuit, and the appliance itself may warrant further testing to check for leakage right at the device.

5 If turning off breakers and switches produces no change in the residual current at the shore-power cord, then the source of the leakage current is another boat plugged into the same shore-power circuit.

6 Try to isolate the problem by unplugging each boat at the dock, one by one (see below).

7 If the reading on your amp clamp goes to zero, you've found the culprit.

It's also likely that the reading may get dramatically lower but not drop to zero because there is a real possibility the source of leakage is more than one boat, and that what you are reading on your meter is the cumulative leakage of multiple faulty boats. To protect your boat, you may need to gain access to several others. Did I mention inconvenience?

Since you can't just turn off the power to everybody's boat without getting permission, I generally notify the marina operator that a serious safety problem exists that needs immediate attention. The manager should immediately prohibit swimming from the docks until the problem is resolved, and if the marina operator is serious about customer service and customer safety, he or she will promptly grant you the permission you require. If you need to make your case more strongly, report your own leakage test results in the context of Table 10-1, which shows the very low levels of current capable of causing death.

Measuring Voltage Potential

Although you can't measure current flow through the water by any simple method, you can measure voltage potential. We are actually more concerned about voltage potential anyhow. In this case, we are concerned about electrical shock hazard and voltage potential applied to human bodies as they swim through the water. The actual current flow through the body will vary depending on the person, and factors such as skin resistance and how long the person has been in the water will determine actual current exposure. All you need to test this is a conventional DVOM set to the VAC scale and an extended lead for the hot terminal. Additionally, a broomstick is quite helpful to make an extension probe for the meter (see the sidebar on page 98). With these tools, follow these steps:

1 Confirm the integrity of the ground with the SureTest tool.

2 Attach the hot lead of the DVOM to the end of the broomstick, leaving the conducting tip exposed.

3 Plug the meter's other lead into the grounding terminal on the shore-power pedestal.

4 With the meter set to VAC, probe the water around the dock and boat looking for a VAC reading.

5 You won't see a solid 120 or 220 VAC reading, but you could easily see 60 to 100 VAC if a problem exists. If that's the case, immediately notify the marina

MAKING AN EXTENSION PROBE

To make an extension probe, you'll need the following supplies:

    broomstick

    crimp-on ring terminal

    several small screws

    square of copper flashing

    extended multimeter lead

    To begin, remove the pointed probe end common to most meter leads and attach the crimp-on ring terminal. The terminal should have a hole in it for the small screw you'll use to attach the copper plate.

    You can get copper flashing at your local hardware store. (Home Depot sells 5-by-7-inch precut pieces that are just right for this job.) The actual size isn't critical, but bigger is better since it makes the lead more sensitive to finding current fields.

    Use small screws to attach the copper square to the lead and secure it to the end of the broomstick handle.

extended lead   meter   black lead (plugs into the grounding terminal on the shore-power pedestal)   copper square (goes in water)

Here's the extended probe rig I made.

operator and all the boatowners in the area—a potentially deadly situation exists. (Note: Your previous residual current readings and SureTest readings are clues to this possibility. Take these clues very seriously; you could save somebody's life.)

AC LEAKAGE FROM DEGRADED INSULATION

The second common cause of residual current is the breakdown of insulation in motor and generator windings and in shore-power isolation transformers. Leakage can occur between adjacent windings and between the windings and case grounds. It can also be influenced by ambient temperature and humidity. However, this breakdown can be difficult to detect because the insulation may still retain *most* of its insulating properties, and the device may continue to function normally even as the leakage is occurring.

    The effectiveness of insulation is a function of its resistance (impedance), measured in ohms (Ω). Materials that are used as insulation of course have very high impedance—so high, in fact, as to be off the scale of the ohmmeter function on a standard DVOM, even

MEGOHMMETERS

Model 380360, Extech Instruments, www.extech.com
Model 1026, AEMC, www.aemc.com

after they have begun losing their effectiveness. This is not surprising, since ohmmeters are typically used to look for failures of conductivity, not of insulation.

Enter the megohmmeter, commonly known as a megger. The megger is used to measure high resistance values in insulation in the mega, or millions-of-ohms, range (1 megohm = 1,000,000 ohms). The model I use (an Extech 380360, which costs just under $200) can read up to 2,000 MΩ, or 2 *billion* ohms. That is *very* high resistance.

Caution: Be extremely careful when using a megger; it can destroy circuitry and cause personal injury or death. As with other instruments we've covered, the megohmmeter sends a signal through the circuit—only in this case, it's anywhere from 250 to 1,000 volts! (In contrast, the typical DVOM set to the ohms scale sends only about 1 or 2 volts through the circuit.) This amount of voltage can do some damage—not only to sensitive electronic control circuitry but to you.

Because of its powerful capabilities, the use of a megger has some limitations and precautions:

- Only use a megger on heavy-gauge wire winding sets.
- Only use a megger when the manufacturer of the equipment specifies the use of one as a test method and provides instructions on how to isolate any circuitry associated with the windings.
- Only perform megohmmeter tests with the equipment workshop manual in

hand, and follow the manufacturer's test procedure and specifications to the letter.

The specification I use regularly is a minimum of 1 megohm, which is based on generally accepted U.S. military and Transport Canada guidelines for ships and large "rotating machinery," as they often refer to generators, and is their minimum requirement.

A megohmmeter test is often timed, as it takes a while for the windings to become "saturated" as the high voltage is applied. In a maintenance situation, technicians may use a megger to periodically test for gradual deterioration of insulation over specified time intervals, so extended tracking and data logging is sometimes necessary. For most recreational boat applications, however, this sort of work is quite uncommon, as the labor costs will typically outweigh the cost of simply replacing components.

So when do I use a megger? If I get higher than normal residual current readings from a boat's shore cord, for example, and I am able to trace the leakage to a particular circuit, I use the megger to further isolate the problem. If the circuit feeds a refrigeration system, then the compressor motor windings are suspect. If excessive leakage is detected from a shore-power-supplied battery charger, its transformer windings may be the problem. If residual current is only noticed when the AC system is being supplied by an onboard generator, then its windings may be at fault. All of these problems may be caused by insulation breakdown, and a megger might be the best tool to confirm or reject that suspicion. Alternatively, many people will use the clamp-on leak tester discussed earlier. If excessive leakage is found, they'll just replace the piece of equipment.

USING A MEGOHMMETER

Meter lead connections and screen displays vary from one brand of megger to another, but there are some commonalities among meters:

- multiple voltage test ranges
- a low ohms scale for checking continuity and resistance values below 200 Ω
- the ability to measure voltage up to 600 VAC
- a "3-minute" lock test function (a common time parameter used by equipment

manufacturers who recommend periodic insulation tests)

It is common practice to select a voltage range that is approximately twice the value of the rating of the appliance or cable being tested:

- For a 120 VAC appliance, use the 250 V scale (or the nearest equivalent on the meter).
- For a 220 VAC appliance, use the 500 V scale.
- For 600 V boat cable, use the 1,000 V scale.

Connecting the meter to the circuit or appliance in question is easy once you've sorted out any isolation requirements and

The Extech 380360 megohmmeter measures resistance up to 2,000 MΩ. It also has a 3-minute lock test function. Also shown is the unit's master test button.

transformer
primary
winding lead

chassis ground
should be
electrically
isolated

Testing for leakage and good ground isolation on a transformer winding set. Notice the meter lead connections at both the transformer and at the meter end. The stud on the motor is the ground. The meter's other lead is clipped to one of the primary winding leads.

other specified criteria, such as temperature. For example, if you are testing cable insulation, make sure the cable is completely disconnected from the circuit it serves at both ends. For ground leakage tests on a tool or appliance, there are no special concerns to address. But for motor circuits, transformer windings, and generator winding sets, it's best to follow the advice of the equipment manual to ensure that you don't do more harm than good. As for temperature concerns, some equipment is tested at normal operating temperatures. In other cases, the equipment manufacturer's test procedure may require you to adjust the effective resistance value up or down based on actual temperature.

Even with a maximum measurement range of 2,000 MΩ, it's not uncommon for a megohmmeter to max out during testing.

Different models indicate this in various ways; on the Extech model 380360, off the scale is indicated by a screen display of 1 followed by blank spaces, a decimal point, and the symbol MΩ (see photo page 102). This simply means you have excellent insulation—it's not an error.

VALUE AND UTILITY

Meggers can be extremely useful, but the range of applications for most marine technicians and boatowners is quite limited, and most technicians do not use them very often. Unless you work on mega- or superyacht-sized vessels, or in an industry where leakage and resistance checks are regularly required, a clamp-on leak tester will supply most of your needs.

To test a piece of coaxial cable for leakage, insert the lead between the core conductor and shield.

A Basic Corrosion Survey

In this chapter, we will look at some of the tools and procedures we use to conduct a basic underwater metal corrosion survey. These procedures typically fall into a specialized subcategory in the marine service world, that of a corrosion specialist. (In fact, one of the certifications that the ABYC offers is for corrosion. Interestingly enough, the ABYC also strongly recommends that those who sit for this certification have the ABYC Electrical Certification as a prerequisite.) Underwater metal corrosion can be caused by failures or improper installations of electrical systems, however, so general marine electricians and dedicated do-it-yourself boatowners should at least be aware of the basics of the subject. Those who wish to delve further should contact the ABYC's Education Department and register for a Corrosion Certification course and exam.

My goal in this chapter is to provide you with enough knowledge to perform a basic corrosion survey, which will include looking for stray DC current caused either by faulty wiring or defective DC appliances. I will also cover the galvanic isolator and how to determine if it is in good working order. Since a galvanic isolator may be an integral part of the boat's AC grounding system, its proper functioning is critical to crew safety.

TOOLS OF THE TRADE

The corrosion specialist uses a variety of arcane equipment, special reference electrodes, conductivity testers, water pH and salinity testers, split-out shore cords, prewired galvanic isolators, extra anodes of different materials, and an absolute plethora of gear to do his or her work.

However, for the marine technician and advanced boatowner, I'm going to focus on two specific tools: a silver/silver chloride reference electrode and an extremely high resolution DC amp clamp, such as the one shown in the photo on page 117. In addition, you will need your standard DVOM.

A special electrode is necessary because (1) we will be measuring extremely small voltage potentials; (2) we will need extremely high levels of sensitivity and precision; and (3) we must be able to repeat the test in a variety of electrolytes (i.e., water of various degrees of salinity). The silver/silver chloride electrode meets all these requirements; it is the most common reference electrode used in marine applications due to its ease of manufacture and its stability over a wide range of temperatures.

A conventional DVOM and a silver/silver chloride reference electrode. This one is made by USFilter (now called Siemens Water Technologies) under the trade name Capac, but the cells are readily available through Mercury engine special tools catalogs or West Marine. The electrodes cost about $85. The Capac unit costs about $250 and is equipped with an extremely long connecting lead for use on larger boats.

Many technicians prefer to use a common zinc electrode as a reference cell, and zinc will work. However, I prefer to use a silver/silver chloride cell because ABYC Standard E-2, Cathodic Protection, bases all its voltage potential values on the silver/silver chloride cell (see the table on page 106). If you choose to use a different reference electrode, the standard does provide conversion tables. In the case of zinc, you would add 1 volt to your meter reading to convert to the values in Standard E-2. Personally, I find this all rather confusing and prefer not to deal with conversions in a working situation.

UNDERWATER METAL CORROSION AND PROTECTION BASICS

A basic corrosion survey tests the electrical potential of your boat's hull (the metal) to answer two questions:

SILVER/SILVER CHLORIDE REFERENCE ELECTRODES

Capac, Siemens Water Technologies, www.usfilter.com
Model 20008, ProMariner, www.promariner.com

1 Is the boat's cathodic protection system (assuming it has one) doing its job?

2 What is the overall condition of the boat's bonding system?

First, let's review the difference between a cathodic protection system and a bonding system. Simply put, a *cathodic protection system* protects a boat against galvanic corrosion through the use of zinc anodes. A *bonding system* protects the boat against stray-current corrosion by electrically tying together all the metals on the boat and then connecting the bonding wires to the boat's common ground point.

Cathodic Protection and Galvanic Corrosion

When two dissimilar metals are immersed in an electrolyte, their differing electrical potentials will produce a voltage difference. Anytime a connection is made between these two metals, an electrical current will flow from the higher-voltage metal (the cathode) to the lower-voltage metal (the anode), leading to the deterioration of the anode. Since the cathode in a galvanic cell does not dissolve, it is also referred to as the most noble metal, while the anode is referred to as the least noble. Note that this relationship is relative; depending on its position in the galvanic series table (see page 106), a metal will be cathodic with respect to metals that have a lower electrical potential but anodic with respect to those with a higher potential.

Applying this to boats and the marine environment, it's clear that since metal fittings and components on boats are made of dissimilar metals and are usually underwater, you have the makings of a galvanic cell on your boat. Add an electrical connection, and galvanic corrosion will result.

Cathodic protection is the answer to basic galvanic corrosion. By installing zinc anodes

on your boat, you provide an anode that will be consumed (sacrificed) to protect the boat's metal fittings and components.

The scenario becomes a bit more complicated when you hook up to dockside power. Now the external circuit becomes the green grounding wire, which basically connects every boat to one another. As a result, you could find your boat becoming the cathodic protection for your neighbor's boat, and all of your zinc anodes will start corroding very quickly. Once they are gone, the next least noble metal on your boat will follow.

The solution to this problem is a galvanic isolator. Basically, this device interrupts the circuit between your boat and the other boats tied into the dockside power. It has a pair of diodes that block low DC voltage, up to 1.5 V, but allow higher AC voltage to pass, which is an important feature because as mentioned in Chapter 10, even low-level AC voltage in the water is a danger to swimmers. Thus while providing a solution to galvanic corrosion, the galvanic isolator also provides protection from dangerous AC voltage.

Bonding and Stray-Current Corrosion

Stray currents can come from faulty electrical circuits or from any situation that provides electrical current with a path that has a lower resistance to ground than the appropriate one back to the battery. For example, if you have a connection sitting in a puddle of water, the water provides a path of lower resistance than the wire going back to the battery. As this current follows its path of least resistance through metal into the water, it will corrode the metal. It doesn't matter if the metal is more anodic or more cathodic (zinc anodes won't work here)—the current will corrode the metal, and the direction of the current flow determines which metal will corrode.

Metals and Alloys	Range of Corrosion Potential (relative to silver/silver chloride half cell; volts)
Magnesium and Magnesium Alloys	−1.60 to −1.63
Zinc	−0.98 to −1.03
Aluminum Alloys	−0.76 to −1.00
Cadmium	−0.70 to −0.73
Mild Steel	−0.60 to −0.71
Wrought Iron	−0.60 to −0.71
Cast Iron	−0.60 to −0.71
13% Chromium Stainless Steel, Type 410 (active in still water)	−0.46 to −0.58
18-8 Stainless Steel, Type 304 (active in still water)	−0.46 to −0.58
Ni-Resist	−0.46 to −0.58
18-8, 3% Mo Stainless Steel, Type 316 (active in still water)	−0.43 to −0.54
78% Ni/13.5% Cr/6% Fe (Inconel) (active in still water)	−0.35 to −0.46
Aluminum Bronze (92% Cu/8% Al)	−0.31 to −0.42
Naval Brass (60% Cu/39% Zn)	−0.30 to −0.40
Yellow Brass (65% Cu/35% Zn)	−0.30 to −0.40
Red Brass (85% Cu/15% Zn)	−0.30 to −0.40
Muntz Metal (60% Cu/40% Zn)	−0.30 to −0.40
Tin	−0.31 to −0.33
Copper	−0.30 to −0.57
50-50 Lead/Tin Solder	−0.28 to −0.37
Admiralty Brass (71% Cu/28% Zn/1% Sn)	−0.28 to −0.36
Aluminum Brass (76% Cu/22% Zn/2% Al)	−0.28 to −0.36
Manganese Bronze (58.5% Cu/39% Zn/1% Sn/1% Fe/0.3% Mn)	−0.27 to −0.34
Silicon Bronze (96% Cu max./0.8% Fe/1.50% Zn/ 2.0% Si/0.75% Mn/1.60% Sn)	−0.26 to −0.29
Bronze Composition G (88% Cu/2% Zn/10% Sn)	−0.24 to −0.31
Bronze Composition M (88% Cu/3% Zn/6.5% Sn/1.5% Pb)	−0.24 to −0.31
13% Chromium Stainless Steel, Type 410 (passive)	−0.26 to −0.35
90% Cu/10% Ni	−0.21 to −0.28
75% Cu/20% Ni/5% Zn	−0.19 to −0.25
Lead	−0.19 to −0.25
70% Cu/30% Ni	−0.18 to −0.23
78% Ni/13.5% Cr/6% Fe (Inconel) (passive)	−0.14 to −0.17
Nickel 200	−0.10 to −0.20
18-8 Stainless Steel, Type 304 (passive)	−0.05 to −0.10
70% Ni/30% Cu Monel 400, K-500	−0.04 to −0.14
18-8, 3% Mo Stainless Steel, Type 316 (passive)	0.00 to −0.10
Titanium	−0.05 to +0.06
Hastelloy C	−0.03 to +0.08
Platinum	+0.19 to +0.25
Graphite	+0.20 to +0.30

Anodic or Least Noble (Active) ↑

Cathodic or Most Noble (Passive) ↓

The galvanic series table. (Reprinted with permission from Boatowner's Mechanical and Electrical Manual, third edition, by Nigel Calder)

A bonding system. (Reprinted with permission from Boatowner's Mechanical and Electrical Manual, *third edition,* by Nigel Calder)

Bonding electrically ties together all major fixed metal items and then connects them to the boat's ground. This has two purposes:

1 It removes the conditions for current to flow. The key to inducing electrical current flow in metals is a difference in voltage potential. If there is no potential difference, there can be no current flow. By electrically tying all of the underwater metals together, we create an electrical circuit where no potential difference exists, therefore no current will flow. No current flow, no corrosion.

2 It removes the possibility of an inappropriate path to ground, such as through the hull or through-hull fittings. Bonding maintains a low-resistance path back to the battery.

In the United States, the bonding approach is commonplace. Underwater metals connected via a bonding system will polarize (see below), and as long as they are connected to an appropriately sized anode—either zinc, aluminum, or magnesium (depending on the location of the vessel)—the anode will become the sacrificial lamb in terms of corrosion. One argument against this approach concerns outside forces, such as current leaks in the water at a dock that can migrate and be distributed via a bonding system inside the boat. So some builders believe isolating the metals is a better approach. Many European boatbuilders take this approach. In the United States, we favor bonding; the ABYC advocates bonding, and I support that position for most (but not all) boats.

In some cases, bonding is just not practical or necessary. For example, small trailerable boats with only a few through-hull fittings aren't really a concern. They can be easily monitored, and if the boat spends most of its life sitting on a trailer, bonding is overkill. For wooden boats, bonding may not be a great choice either; faulty cathodic protection systems on wooden hulls can wreak havoc with the wooden structure of the boat.

CONDUCTING A CORROSION SURVEY

Most corrosion surveys are performed because the boatowner is having a problem: either excessive corrosion or, more commonly, rapid anode consumption. Troubleshooting the problem requires first gathering some data: water temperature, current, depth, and salinity; boat usage; and the presence of new boats in the area. These can all impact how long an anode lasts, and you (or, in complex cases, a corrosion specialist) will spend quite a bit of time researching changed conditions when the complaint is rapid anode consumption.

For the purposes of this book, we'll confine ourselves to learning how to answer a few basic questions:

- Is the boat cathodically over- or under-protected?
- Is the boat equipped with a galvanic isolator? If so, is it functioning properly?
- Is there any electrical current leakage into the bilge water or the bonding system? If so, which circuit is causing this induced current?

We'll begin by confirming the calibration of the DVOM; it should be correct to within less than 50 millivolts (mV). We will be measuring very small values, and sensitivity is quite important.

Here's how to check calibration using a zinc anode and a silver/silver chloride reference electrode:

1 Using a small-engine pencil zinc anode as your test piece, attach it to your DVOM's positive lead.

2 Connect the silver/silver chloride reference electrode to the DVOM's "com" or negative jack (top photo opposite).

3 Set the meter to DC volts and submerge the zinc and reference electrode in salt water. Be sure the electrodes are not touching the bottom; 1 foot or so below the surface is just right. The reference cell and the zinc anode should be within several feet of each other.

4 The meter should read approximately −1 volt DC (bottom photo opposite), indicating that the meter is calibrated correctly. A reading of 50 mV or so one way or the other indicates the zinc alloy varies slightly, which changes its actual voltage potential.

5 If your meter readings are not approximately −1 VDC, try another anode; the one you are using may not be zinc, but rather magnesium or an aluminum alloy.

6 If a new anode doesn't help, see if another DVOM makes a difference. If it does, then your meter is at fault.

It's important to note that the voltage potentials given here assume you are working in a saltwater environment. In fresh water, due to its relatively low conductivity, voltage potentials will be much lower. For example, a boat in salt water that might give a hull potential reading of approximately –550 mV to –900 mV would only give a reading of –300 mV to –400 mV in fresh water. Of course, the reading you get will depend on the actual conductivity of the water the boat is floating in.

Corrosion specialists that work in freshwater environments regularly use a copper sulfate reference electrode to help offset this conductivity disparity. A copper sulfate electrode typically reads about –100 mV higher than a silver/silver chloride electrode, keeping actual readings more in line with acceptable protection levels as described in ABYC Standard E-2. A good source for copper sulfate reference electrodes is the M. C. Miller Company, www.mcmiller.com.

Follow these steps to measure a boat's basic hull potential:

1 Ensure all normal dock connections are in place (e.g., shore power, cable TV, etc.)

2 Plug the reference electrode into your DVOM's "com" or negative jack, and submerge it in the water; don't let the electrode touch bottom—about mid-draft for the boat works here.

3 Touch the DVOM's positive lead to either the boat engine's negative terminal or the primary grounding bus bar behind the main electrical distribution panel.

4 The voltage reading on the DVOM represents the hull's basic potential. It should be a negative number similar to those shown in the galvanic series table (see page 106).

The actual reading you get will be an average of the potentials of all of the underwater metals connected to the grounding system as compared to the reference electrode. This "average" is what happens once metals are connected

To check the test instrument prior to conducting an underwater corrosion survey in salt water, clip a pencil zinc to the meter's positive lead and attach a silver/silver chloride reference electrode to the negative lead as shown.

(via a bonding system) and submerged in an electrolyte. This is an electrochemical process called *polarization*, in which the inherent potential of the various bits of submerged metal changes, and they begin to equalize, or come to a common average

With both leads in the water off the side of the dock (just outside the photo), the meter should read approximately –1 VDC, as shown.

potential. The idea is to make sure that the potential reading you get with an anode (or anodes) attached to the system is at least −200 mV greater (more negative) than the average reading without an anode attached. Actual numbers will vary depending upon the number and surface area of the anodes, as well as the specific alloy the anodes are made of.

5 Compare your readings to the recommended values in ABYC Standard E-2, shown opposite, to see if your boat is adequately protected.

The reading in the bottom photo opposite is slightly more negative than the ABYC table suggests. This is not unusual and represents no problem in the case of a fiberglass boat. All that is implied is that the anode surface area is slightly greater than needed, and since the boat being tested just recently had its anodes replaced, this is to be expected. As the anodes are depleted, the exposed surface area will be reduced and the voltage potential will become more positive. In fact, I generally recommend adjusting the anode area to give a slightly more negative reading than the ABYC table suggests as a starting point, knowing that in a relatively short period the numbers will fall within the prescribed range. All this does in effect is increase the effective service life of the anodes by a few weeks or a month.

What if your readings do not fall within the ranges recommended by the ABYC E-2 standard?

If the readings are less negative than recommended, your boat does not have adequate anode surface area exposed to the electrolyte in which it is floating. There are two explanations for this scenario:

1 If the anodes have been doing their job for the better part of a boating season, then the situation is normal. The decrease in surface area means the zinc material has been sacrificing itself (eroding) to protect

Testing your boat's corrosion potential. (Reprinted with permission from Boatowner's Illustrated Electrical Handbook, second edition, by Charlie Wing)

Touch the meter's negative ("com") lead to the battery negative return, or engine negative terminal, as shown.

A typical meter reading, –0.955 mV, for a fiberglass boat with bronze running gear and a typical array of bronze through-hull fittings.

your valuable metal components. It's time to replace them.

2 If the anodes were replaced recently, then the rapid loss of material means that you may never have had enough anode material to begin with, or that your anodes are protecting other boats on the dock due to a lack of a galvanic isolator, or that AC or DC stray currents may be at work.

If the readings are more negative than recommended, there are two possibilities:

1 It is possible someone has inadvertently installed magnesium anodes. These are used in fresh water and should never be used in a saltwater environment as their negative potential is high enough to create an overprotection situation that can be very damaging to aluminum hulls and outdrives. (Aluminum is unique among the common metals used in marine construction in that it can be damaged by either under- or overprotection from a cathodic protection system.)

2 There is a DC current leak, either through a connected fitting on the boat or from an adjacent boat—via the water—that has a DC stray-current problem. (We'll look at how to check for this shortly.) If the current leak is coming from someone else's boat, call a corrosion specialist as it's going to take some serious tracking down to isolate.

HULL MATERIAL	MILLIVOLT RANGE
Fiberglass	–550 to –900
Wood	–550 to –600
Aluminum	–900 to –1100
Steel	–800 to –1050
Non-metallic w/Aluminum Drives	–900 to –1050

The recommended range of cathodic protection from ABYC Standard E-2. (Courtesy ABYC)

Let's assume the reading you get is within the prescribed range. You're not done yet:

1 With your meter still connected and the reference electrode still submerged, remove the electrical connections from the boat to the dock one at a time and watch for a change in the meter readings.

2 If the reading gets more negative, then there was leakage from the boat to the dock through the shore cord you just disconnected. The subject boat may be helping to protect other boats at the dock with its anodes.

3 If the reading gets less negative when disconnected, there may be a DC stray-current problem on board the boat. Further checking will confirm or deny this possibility.

4 If there is no change in the reading with the boat plugged in versus not plugged in, two possibilities exist:

• The most likely reason is the boat has an isolation transformer or galvanic isolator that is doing its job.
• There is a no-ground situation at the dock—a potentially deadly situation. It is essential that you check the integrity of the dock ground with the SureTest tool (Chapter 4).

Next, let's test the integrity of the bonding system on board the boat. We'll use the DVOM to check the hull potentials:

1 First, perform this test with the engine(s) off.

2 Find all the metal objects in the bilge that are attached to the bonding system.

3 With your reference electrode still in the water near the boat, use your positive meter lead and check the

potential at each bonded metal fitting one by one. The readings should be the same, or within about 50 mV of the basic hull potential; otherwise, the possibilities are:

• If a reading is more positive than the basic hull potential, a poor connection in that leg of the circuit is indicated, and it should be tracked down and repaired.
• If a reading is dramatically more positive or negative than the basic reading, it may indicate DC stray current, which should be remedied.

4 Now repeat the test with the engines running. Dramatic changes in the readings may indicate DC stray-current problems induced by something electrical that is running on the engine, typically an alternator.

Checking potential at a seacock (touch the lead to any metal part). The meter reading is within 1 mV of the overall hull potential shown in the bottom photo on page 111, indicating that all is well.

Finally, we need to check the dock's potential. For this, we'll move to the shore-power pedestal:

1   With the reference electrode still in the water, plug the DVOM's positive lead into the ground lug on the dock receptacle.

2   **Caution: A shock hazard exists. Be sure you have positively identified the ground lug at the pedestal, not the hot lead.** On 30 amp service, this is typically the notched or L-shaped lug. On 50 amp service, it is a shell connection, as shown in the photo. See the illustrations for other common shore-power terminal configurations.

3   Most docks will read between –400 mV and –800 mV. A reading outside that

The shell is the ground connection on a typical 50 amp shore-power receptacle.

120 Volts

15A, 120V
Straight blade
2 pole, 3 wire

Receptacle

Plug

20A, 120V
Straight blade
2 pole, 3 wire

Receptacle

Plug

20A, 120V
Locking
2 pole, 3 wire

Receptacle

Plug

30A, 120V
Locking
2 pole, 3 wire

Receptacle

Plug

50A, 120V
Locking
2 pole, 3 wire

Receptacle

Plug

240 and 208 Volts

50A, 120/240V
Locking
3 pole, 4 wire

Receptacle

Plug

30A, 120/208V 3ØY
Locking
4 pole, 5 wire

Receptacle

Plug

100A, 120/240V
Pin and sleeve
3 pole, 4 wire

Receptacle

Plug

100A, 120/208V 3ØY
Pin and sleeve
4 pole, 5 wire

Receptacle

Plug

Dock and shore-cord terminal configurations, from ABYC Standard E-11. (Reprinted with permission from Boatowner's Illustrated Electrical Handbook, second edition, by Charlie Wing)

range could indicate DC stray current migrating through the dock ground system. This will require investigation.

4 Assuming the dock's basic potential is between –400 mV and –800 mV, compare that number to the boat's basic potential both when plugged into the dock and when unplugged.

5 The basic hull potential with the boat plugged in should fall between the basic dock potential and the boat's unplugged potential. Although the exact reading for the unplugged boat may not be the ideal value for the specific boat you are checking, it may be acceptable if anode consumption is within the acceptable service limits of three to ten months for a typical recreational boat.

Proper electrical location of a boat's galvanic isolator, from ABYC Standard A-28. (Courtesy ABYC)

Galvanic Isolators

As mentioned above, each boat plugged in at a dock is helping to protect all the other boats at the dock. They are all married to one another via the green wire. For this reason, I strongly recommend a galvanic isolator for any boat that spends its time plugged in at a dock.

The proper location for the galvanic isolator is in series with the green grounding conductor, and as close to the AC inlet on the boat as possible. Install it so that it can't be bypassed electrically, as shown in the illustration opposite.

You may need to test the galvanic isolator to ensure that it is functioning properly. An ABYC-compliant isolator is made up of two sets of opposed diodes and a parallel-connected capacitor (exclusive of any status-monitoring circuitry that may be associated with the device), as shown in the detail in the illustration. You must confirm continuity for current flow in both directions to rule out a case short to ground of one or more of the diodes and to ensure the capacitor is functional.

Testing a Galvanic Isolator

In most cases, this test is a three-part process, as outlined below.

Step 1. Check the continuity of the diode set in one direction:

1 Connect your DVOM leads to the terminals where the boat's green wires attach. It doesn't matter which lead connects to which terminal; you will be checking continuity in two directions through the device.

2 Set the DVOM to the "diode" test function.

3 With the leads connected, watch as the meter's reading gradually increases, showing that the battery in your DVOM is charging the capacitor inside the galvanic isolator.

4 If you get an immediate continuity reading through the isolator and no

Checking the continuity of the diode set in one direction. A gradual rise in the reading indicates the charging of the capacitor from the battery of the DVOM. Once charged, the meter reading will stabilize. The reading is essentially the amount of voltage drop across the isolator—typically around 0.9 volt.

gradual increase in your reading, then the isolator is not equipped with a capacitor and is not compliant with current ABYC standards.

5 Assuming the isolator has a capacitor, wait until the numbers on the DVOM stop increasing (this may take a few minutes). Make a note of the number.

Step 2. Check the continuity of the diode set in the opposite direction:

1 Disconnect the meter leads and "flash" the two terminals on the isolator to discharge the capacitor.

2 Reverse the meter's leads from the previous connection and repeat the Step 1 test above.

3 If the capacitor is in good condition, the meter reading should stop within 10% to 15% of the value from Step 1.

If you get a no-continuity reading from either test, the isolator diode sets have failed to

Testing the continuity in the opposite direction. First "flash" the capacitor by shorting it from one side to the other with a jumper lead, and reverse the DVOM leads. Again the meter reading will rise gradually as the capacitor charges. Once the meter reading stabilizes, compare it to the first reading.

Testing for a diode short circuit to ground. Connect one meter lead to either of the green wire terminals entering the device and the other meter lead to the case of the isolator. Continuity indicates a short circuit, and the isolator must be replaced.

an open circuit and immediate replacement is needed.

Step 3. Determine if a short circuit to ground exists:

1 Connect one meter lead to either of the green wire terminals and the other lead to the isolator case (you may need to scrape a bit of paint off the surface to acquire a good connection).

2 Any continuity reading indicates a short circuit to ground; replace the isolator immediately.

3 A large variation in the comparative readings on the capacitor charge would also indicate replacement, but I have never seen this happen.

Due to the inherent voltage drop that occurs across a diode—typically 0.6 V to 0.7 V—

galvanic isolators can be used to effectively block galvanic currents up to about 1.4 V. By putting two diodes in series, you can block 1.4 VDC. Galvanic isolators with capacitors are designed to offer extra protection for the diodes in the event of a continual, low-level AC leakage current running through the grounding wire. This is not uncommon, as discussed in Chapter 10. AC passes through a capacitor; DC will not. The capacitor provides a path for any AC leakage, or fault current for that matter, to bypass the diodes, but still effectively blocks DC current flow. When we discuss galvanic current, we are referring to direct current.

Tracking Stray Current

So far we've only touched briefly on stray current, and evidence of its existence is based on various hull or dock potential readings with the silver/silver chloride reference electrode (excessively high negative meter readings can indicate DC stray current). Stray current is what I like to refer to as battery-level voltage and current, as compared to the galvanic current we have been talking about up to this point— voltages less than 1.4 VDC.

HIGH-RESOLUTION DC AMP CLAMPS

Model 380942, Extech Instruments, www.extech.com
Model 100-LCD-200-3/4" Sea Clip, Swain Meter Company, www.swainmeter.com

Battery-potential stray current is another matter, and it can cause profound metal corrosion in as little as 24 hours! If rapid corrosion is noted and confirmed, it's important to take quick action to prevent further damage to through-hulls and other bonded underwater metals. This procedure is best done with a helper, as jumping in and out of the bilge can get a bit tiresome.

Identifying Leakage Current with a DC Amp Clamp

To detect the low-level leakage resulting from bonding and grounding systems, you need a high-resolution DC amp clamp, such as the Extech 380942. If you can't measure down to 0.1 mA AC and 1.0 mA DC, you may not be able to detect leakage current that is strong enough to cause profound corrosion and, as discussed earlier, be potentially lethal to swimmers around the boat.

If you suspect that DC stray current may be causing problems on board your boat, find the source using process of elimination (see illustration page 118):

1 Start the test with the engine off.

2 Attach the amp clamp around the bonding conductor.

3 Activate the circuits on the DC panelboard one at a time.

4 If you suddenly get a reading on the amp clamp, you have identified the problem circuit.

5 Use a TDR to pinpoint the location of the leak (Chapter 3). The TDR is the best tool to find the location of a short circuit to ground within a circuit.

6 Repeat the test with the engine running to check a faulty alternator or starter motor.

Identifying Leakage Current with a DVOM

It is also important to check the bonding system for current flow at several locations. It is possible for leakage current from different locations, such as bilge pumps and float switches, to "dump" itself out through the bottom of the

The high-resolution Extech 380942 amp clamp can measure down to 0.1 mA AC and 1.0 mA DC.

A typical bonding system, showing locations to test with the amp clamp while activating DC circuits, to see if any stray current exists. (Reprinted with permission from Boatowner's Mechanical and Electrical Manual, third edition, by Nigel Calder)

boat at only one through-hull point, bypassing the others. Perform this test as follows:

1 Partially flood the bilge with salt water.

2 Set your DVOM to DC volts.

3 Attach the negative or "com" lead to the engine or battery ground (negative terminal).

4 Activate the suspect bilge pump and use the positive lead as a probe.

5 Tracking through the bilge water, slowly home in on the float switch or pump, as shown in the illustrations opposite.

6 If you start to get a voltage reading on the meter, you've found a leaking component.

7 Replace the component immediately as it will cause rapid corrosion of metal parts sharing the same bilge water.

8 Another possible source of the voltage reading could be a stray, unterminated conductor lying in the bilge or a connected conductor with bad insulation.

This chapter has touched on some of the classic and routine tests you or a marine electrician can use to confirm that your boat's electrical system, which is tied into the cathodic protection system, is in good order. These relatively straightforward tests will also help you determine if you need to call in a certified corrosion specialist.

A leaking seal on a float switch will result in stray-current corrosion at the positive switch terminal. This will eat the positive cable off the terminal, disabling the switch and pump.

⊖ from battery ⊕

current is fed from the positive terminal on the switch to the other terminal through the water

⊖
⊕

water in bilge

water penetrates switch seal

A radio grounded to a through hull. The regular ground wire has a loose connection at the negative bus. Current flows to ground through the seacock.

battery isolation switch

+   −

battery

loose or broken connection

electric anchor windlass

common ground point

radio ground plate

+ 2.0 volts

Stray currents eat away hardware.

stray currents

Exposed connections in damp area of the boat provide a path to ground.

bilge pump

seacock

Internal leak in bilge pump runs to ground through bilge water.

Dampness provides a path to ground.

2.0 volts

Inadequate ground wire leads to voltage drop. Current finds a more direct path to ground via damp and seacock.

Use a standard DVOM to track DC stray current through bilge water and home in on electrically "leaky" devices like pumps and float switches, and to find active DC positive conductors lying in bilge water. (Reprinted with permission from Boatowner's Mechanical and Electrical Manual, third edition, by Nigel Calder)

Grounding Systems

L et's delve a little deeper into the vital but often ignored subject of grounding systems. Depending upon a boat's equipment list, a grounding system will serve as many as four primary needs:

1 Protect individuals from electrical shock (see Chapter 10).

2 Minimize corrosion (see Chapter 11).

3 Provide lightning protection.

4 Minimize RFI and/or provide a radio antenna counterpoise.

In this chapter we'll look at lightning protection and the basic radio frequency ground plate. In Chapter 16 we'll look at what happens when we tie in an SSB radio counterpoise, which is basically a slightly more elaborate grounding system that serves as half of the radio's antenna system.

LIGHTNING PROTECTION

Lightning is both powerful and mysterious. We have been observing it for centuries, devoting much study and research to learning what it is, how to control it, and how to minimize its powerful effects. That said, and in spite of all our efforts, the best lightning protection system (LPS) really can't guarantee personal safety or protection from equipment damage. I tell my seminar attendees that lightning is truly an act of God, and you can only do so much to insure yourself against a strike; after that, you can just hope—or pray—for the best!

Most of what we know about lightning protection has been figured out by carefully observing the damage after a strike and drawing conclusions from these observations. The stuff is just too hard to catch and experiment with!

So what have we learned from these observations as they relate to boat installations? It can be narrowed down to a few key points, which are compiled in ABYC Technical Information Report TE-4 (formerly Standard E-4), Lightning Protection. Below I've adapted some of the basic guidelines provided in that report for installing an LPS or modifying an existing grounding system:

- Lightning is an electrical discharge, and moves due to a difference in electrical potential. Like all electricity, it's trying to get to Mother Earth, or ground.
- Once "captured" by an electrical conductor, the conductor should provide as straight and direct a route as possible to ground, and have the ability to handle the high voltage and current present.

- The exposed conductor ground plate, or the cumulative total area of multiple through-hull fittings and other exposed underwater metal objects tied into the grounding system, needs to be of sufficient size to ensure a low-resistance electrical connection to earth ground. The ABYC recommends at least 1 square foot of surface area in salt water, which will generally work for everything but SSB. SSB manufacturers generally recommend at least 1 square meter or yard of exposed surface area. Since fresh water is less electrically conductive than salt water, several square feet of surface area should be used for lightning protection.

- To reduce the risk of the lightning "side flashing" as it seeks a path to ground, heavy metal objects in proximity to a lightning conductor should be electrically bonded to the lightning conductor. Also, lightning tends to jump out of "corners" created in the lightning ground wiring system, so minimize sharp bends.

The Elements of a Lightning Protection System

The basic elements of an LPS are shown in the illustration. The primary conductor (also called a down conductor) from the masthead must have the equivalent conductivity of a piece of 4 AWG (21 mm²) wire. You can achieve this by using an aluminum spar as the conductor. Carbon fiber spars are considered nonconductive for this purpose, so 4 AWG wire needs to be added. At the upper end, the conductor must terminate at what the ABYC technical report calls an air terminal—in other words, a lightning rod. The air terminal should be the highest point on the mast by at least 6 inches (150 mm).

The down conductor must be tied directly to a grounding plate (or its equivalent) of at

The essential elements of a lightning protection system. (Reprinted with permission from Boatowner's Illustrated Electrical Handbook, second edition, by Charlie Wing)

least 1 square foot in surface area. You can achieve equivalency in exposed surface area by tying in other underwater metal objects such as rudders, metal keel ballasts, through-hull fittings, and the like to the system, but you should never connect the down conductor directly to a through-hull fitting. If you do, and lightning strikes your boat, the strike may blow the fitting out through the bottom of the boat. A tinned copper distribution bus bar makes a good link for the system if multiple underwater metal objects are connected together via the grounding circuit inside the boat, or make up the entire grounding system connection to the seawater.

All of the tie-ins must use—as a minimum— 6 AWG (13 mm²) cabling. This in itself creates a problem on some boats that already have a bonding system in place, because the standards for that system dictate the use of either copper strapping or minimum 8 AWG (8 mm²) wire. Consequently, if you choose to install an LPS, you would have to upgrade the bonding system wiring. Many people find this to be a tough decision, as it is almost impossible to perform any kind of reliable cost/benefit analysis for installing the system. My view is that if you are going to spend any time offshore, or live in an area of the world that is statistically prone to lightning activity, the scale gets tipped in favor of installing an LPS.

Although more exposed metal surface area is always better, the electrical conductivity of the various underwater metal components might be a concern in some cases. Lead, for example, is roughly a tenth as conductive as copper, so even a good-sized lead ballast keel may not provide sufficient conductivity. Iron, in contrast, is approximately a fifth as conductive as copper, and even bronze alloys that contain copper are only about a third as conductive as pure annealed or hard-drawn copper. The best ground plate for an LPS is clearly a pure copper plate. But even with a copper ground plate in place, it can be worthwhile to increase the total exposed surface area by connecting all available through-hulls (typically of a bronze alloy), stainless rudders, and even lead keel ballasts.

Most of these metal components installed on or through the hull often have a coat of antifouling paint on them. In theory, this insulates the metals from the seawater, but in practice it's not enough to make much of a difference. When we consider the potentials involved with lightning current, it's going to take a better insulator than a coat of paint to hold it back!

Finally, tie in all large metal objects in proximity to any of the cabling included in the LPS, using 6 AWG cabling. (ABYC TE-4 states that any large metal object within 6 feet of a conductor must be connected to the system.) On a typical recreational boat, this almost always means that all metal objects must be tied in, including fuel tanks, engines, and generators. The objective is to minimize the risk of *side flashing*, in which lightning jumps out of the conductor to the metal object in search of a better path to ground (even if there isn't one).

Lightning Protection for Electronic Equipment

What, if anything, does an LPS do to protect sensitive electronic equipment? This is a common question among boatowners who are thinking about installing or upgrading an LPS. The simple answer is, nothing. The only way to really ensure that electronic equipment remains unaffected by a lightning strike is to completely isolate it from the entire boat's electrical system. Since this is generally impractical, it's safe to say that you should suspect the integrity of all electronic equipment after a lightning strike, and carefully check all of its operational characteristics. It's important to note that LPS standards focus on personal safety and protection of a vessel's hull below the waterline, nothing more.

GROUNDING TO MINIMIZE RFI EFFECTS

Electrical noise, which is essentially synonymous with RFI, is interesting stuff in that it can be transferred (radiated) through the air or through the wiring on board a boat. As discussed in Chapter 9, one way to minimize its effects is to create appropriate zones of separation between RFI sources and equipment that may be sensitive to its effects. In the discussion in Chapter 9, however, we were focusing on radiated noise. In this section, we will discuss how to minimize the effects of noise transmitted through the wiring on the boat via proper grounding techniques.

Inductance is a property of a conductor or coil that determines how much voltage will be induced in it by a change in current. *Capacitance*, which is the ratio of the electric charge transferred from one to the other of a pair of conductors to the resulting potential difference between them, exists between any two conductors insulated from each other. In many cases, the two conductors may be carrying equal but opposite electrical charges. The difference between them can be described as a difference in potential, which, as discussed earlier, is synonymous with voltage.

Any wire or conductor that has electrical current flowing through it will have a magnetic field around it, and cabling that runs through wire chases and conduits is a prime source of RFI problems. Depending upon its frequency, RFI traveling down a given conductor may, quite easily, be able to couple with a parallel run conductor both inductively and capacitively.

Both inductance and capacitance can cause a number of electrical problems on board a modern boat, including faulty readings on electronic engine instrumentation and even the failure of an electronically controlled engine. The best way to deal with this issue is to establish a separation distance between the conductors feeding these devices, but proper grounding can also help.

Electrically shielded conductors are common on some systems, such as network wiring, and the shielding will often take care of the problem, but not always. Another approach is

the use of "twisted pair" wiring, in which two or more conductors are twisted together for their entire length. Both techniques are widely used by electronics vendors and on systems such as Mercury Marine's SmartCraft network and Teleflex's i6000 control system. This is yet another example of the blurring of the line between the electrical and electronics sides of the boat.

Noise generated by AC devices can be broken down into two categories: differential mode and common mode. Differential mode noise resides in hot (or ungrounded) and neutral (grounded) conductors, both of which carry current. Common mode noise, on the other hand, describes noise that exists between either the hot and the grounding conductors or the neutral and grounding conductors. In either case, the noise can easily manifest itself as a hum emitted by audio equipment or as flickering or diagonal lines on a television screen. Equipment vendors use filtering systems to help control this, with some approaches being more effective than others. One of the ways to further minimize the noise is the use of proper grounding techniques, such as specialized capacitor-based circuitry. The operative word here is "proper." It's possible, if done incorrectly, to make matters worse.

Ground Loops and the Antenna Effect

A *ground loop* is a situation where there is more than one ground connection between two (or more) pieces of equipment. These duplicate ground paths create a loop antenna that will pick up interference currents. The inherent resistance in the cable leads transforms the current into voltage fluctuations. Subtle induced voltages migrate through the ground loop system, making the electrical potential within unstable and allowing desirable signals and noise to ride together. If the desired signal is from an audio system, the

This typical equipment installation (top) has an unintentional ground loop that will very likely pick up interference. The ground loop above can be easily eliminated with the simple reconfiguration shown (bottom) and still be in compliance with NMEA recommendations.

result is audible noise, or static. In video or satellite TV systems, the audio may be affected and screen "static" may also occur. In computer or navigation equipment, the noise may cause operational problems with the gear.

The NMEA's installation standard recommends connecting a networking cable shield only at one end of the shield. The standard calls for the shield to be connected at the talker end of networked devices (i.e., the data delivery side, as opposed to the listener end, which might simply be a display screen.)

COMPARING ABYC AND NMEA GROUNDING STANDARDS

In the field, no troubleshooter is in a position to determine how equipment is designed internally, at least as far as its grounding system is concerned, or how much inherent noise the device may be generating due to design factors. Therefore, it's best to follow accepted industry standards and specific manufacturer recommendations for electronics installations.

ABYC Standard E-11, AC & DC Electrical Systems on Boats, addresses grounding issues, as does NMEA's equipment installation standard. From the ABYC's perspective, case grounding of equipment is one of the most important considerations for a trouble-free installation. Another is the single-point connection between the DC and AC grounding systems on the boat. The ABYC standards for grounding are summarized in the illustration on page 126, which I frequently rely upon for training purposes.

Keep in mind that this diagram reflects grounding needs *as they apply to lightning protection, shock hazard protection, and corrosion protection* more than to radio/electronic equipment needs and the elimination of electronically induced "noise" (although a radio ground plate is shown). Compare it closely to the illustration on page 127 from ABYC Standard A-31, Battery Chargers and Inverters, and notice what is in effect a ground loop between the AC and DC sides of the battery charger.

The grounding wire on the DC side of the battery charger circuit may seem redundant, but in the event of a case short with the DC positive conductor, it is vital for providing a conductor capable of carrying high current back to the power source to ensure that the fuse or circuit breaker in the DC positive line will trip. In many cases, without this comparatively large conductor, the current would tend to flow down the AC grounding conductor. In all probability, this small conductor will not have sufficient amperage-handling capabilities, so it will overheat and possibly cause a fire.

The reason the ABYC standards don't get too concerned over this loop is based on the premise that the DC grounding conductor is

not normally carrying any current; it's merely in place in case the worst happens. This raises an important point about ground loops, however: they are never a problem unless electrical current is flowing through them. But from the troubleshooter's point of view, that *is* the problem, because the current needed to make electronic noise can be minute—too small to pose a safety problem, but large enough to affect the performance of some electronic equipment.

The NMEA standards are a bit stricter in some cases, and less strict in others. For example, ABYC E-11 states that grounding conductors can be no smaller than one wire gauge size smaller than the DC power feed conductor to the device in question, and in no case smaller than 16 AWG (1 mm²) in size. With AC equipment, the conductors are typically the same size, as the power feeds to AC devices are virtually always installed with triplex cable with equally sized conductors sheathed in the outer jacket. In contrast, NMEA recommends using no smaller than 12 AWG (3 mm²) in any case for grounding purposes and prefers the use of 8 AWG (8 mm²).

The intent of the NMEA grounding standard is also telling, and reinforces the difference between ABYC and NMEA interests. To quote from the NMEA standard: "This section details the recommended standards and practices for installation of grounding systems intended to support electronic equipment on vessels." The NMEA is particularly focused on the proper operation of electronic equipment in a general sense. The ABYC, in contrast, is only interested in the proper operation of equipment if its faulty operation could cause a safety hazard to people on board the boat. It's also important to keep in mind the note that follows the above statement from the NMEA standard: "These are general recommendations. Specific and more stringent requirements may

A typical DC negative system and DC grounding system, adapted from ABYC Standard E-11, Figure 18. (Reprinted with permission from Boatowner's Illustrated Electrical Handbook, second edition, by Charlie Wing)

Battery Charger Installation

Notice the differences between the ABYC standards for grounding in a battery charger installation per ABYC Standard A-31 (shown here), and the more general installation standards for lightning, shock, and corrosion protection as stated in Standard E-11 and shown in the previous illustration. If you study these diagrams carefully, you'll notice that in the Standard E-11 drawing, each component connected to the grounding system has only one wire linking it to ground. No loops are shown. But in the case of the Standard A-31 drawing, a loop is created between the AC and DC sides of the grounding system (highlighted area). (Reprinted with permission from Boatowner's Mechanical and Electrical Manual, third edition, by Nigel Calder)

be specified by the specific manufacturer of an electronic device." So, what *are* the NMEA's general recommendations? I've adapted them as follows:

- All metal-cased electronics equipment shall have a case ground tied to the boat's DC grounding system with at least 12 AWG cable, but preferably 8 AWG.
- All metal structures mounted to the vessel, such as tuna towers, radar arches, and outriggers, are required to be connected to the boat's DC grounding systems by at least two conductors with a minimum

size of 6 AWG. Further, these grounding links are required to be labeled "Ground Connection, Do Not Disconnect."

- Finally, the NMEA standard describes RF grounding quite specifically as a common ground used to reduce stray radio frequency noise and "set" all cases of electronic equipment at ground potential, free from noise sources that may be introduced from being connected to the DC or AC ground buses.

This last statement is where things may get a bit confusing. The standard goes on: "The

RF ground bus shall be connected to the DC ground, and must be directly connected to the engine ground and through-hull ground plate for the specific purpose of reducing RF noise on board." Herein lies the confusion. What is being called for is a dedicated RF ground bus that may be tied into the rest of the boat's grounding buses (AC and DC) and also at the "engine negative terminal" as the ABYC describes it. So the question becomes, is it isolated from everything else or is it connected?

This is a case where a picture says a thousand words. Use the illustration on page 126 as a guide for how these connections should be made, and you will be compliant, not only with ABYC E-11, but also with the NMEA electronics installation standard as it applies to grounding.

If your equipment is installed in accordance with the ABYC recommendations as shown in that illustration, and you still experience noise-related issues, then the solution is an electronics grounding system that is completely isolated from the rest of the onboard system. You will still be in compliance with the ABYC standard, as long as the rest of the components shown in the illustration are wired as shown.

That said, I've rarely seen a situation where the configuration in the illustration caused any problems with equipment operation. In my opinion, the difference between the NMEA statements and the ABYC diagram is one of semantics rather than a clear-cut difference. The NMEA does, after all, reference ABYC E-11

as the guide for equipment wiring within its installation standards.

TROUBLESHOOTING REDUX

Before we move on to Part 2, in which we'll look at electronics installations, I'd like to reemphasize what I feel is crucial to effective electrical systems troubleshooting.

It's important to look at your work and your routines carefully, think about what you do most frequently, and about what you *could* do if you had access to a given piece of gear. If you're a professional marine technician, consider how it might help your bottom line; if you're a boatowner, consider the money you'll spend and the time you'll have to invest to learn how to use a new device, and weigh those against the limited usage to which you, as the technician responsible for a single boat, will probably put it. Go through a cost/benefit analysis to determine which devices make the most sense for you to purchase.

Whether you're a professional or an amateur, don't immediately run out and buy every instrument we've discussed in Part 1. By themselves, they won't automatically make you a better electrician or a smarter boatowner. But if you follow some of my recommendations and suggestions, and attain a clear understanding of the capabilities and the applications of each new instrument before you consider buying the next one, you will have mastered new, useful skills and elevated yourself to a higher level of competency.

MARINE ELECTRONICS INSTALLATION AND TROUBLESHOOTING

The foremost standards group in North America that deals with marine electronics installations is the National Marine Electronics Association (NMEA), which developed its standards, in part, as a response to the cries for help coming from its members—electronics equipment vendors. Why the distress? Many of the performance issues related to electronics equipment aboard boats were the result of shoddy or incorrect installation practices.

My contribution to the dialogue is Part 2 of this book. Here's a brief overview:

- Chapter 13—general issues related to equipment installation.
- Chapter 14—installation and troubleshooting of communications systems (VHF, SSB, and satellite), position-finding systems (GPS), and radar.
- Chapter 15—installation and troubleshooting of supplementary equipment, such as depth sounders, wind instruments, and autopilots.
- Chapter 16—coaxial cables and antennas.
- Chapter 17—onboard networks.

Electronic Equipment Installation Guidelines

My goal in this chapter is to provide an overview of the issues addressed in the NMEA installation standards as well as in the ABYC standards referenced by the NMEA, and to discuss some general material that is not specifically addressed by either organization, including some installation guidelines.

From our point of view as equipment installers and installation troubleshooters, we'll encounter more commonalities than differences between the various makes and models of equipment. For example, installing and troubleshooting a Northstar GPS isn't much different from installing and troubleshooting a Furuno GPS unit.

ERGONOMICS

Although not specifically mentioned in the NMEA or ABYC standards, one of the issues to be considered as part of any equipment installation is the human interface. In the case of marine electronic equipment, such as radars, chartplotters, VHF radios, and fishfinders, the ability to see the display and work the controls easily, especially while the boat is underway, is important. The general term I'll use to describe all of this is ergonomics.

Ergonomics is defined in my Funk & Wagnall's dictionary as "the study of the relationship between man and his working environment, with special reference to anatomical, physiological, and psychological factors; human engineering." So, when considering the installation location for electronic equipment on board a boat, you have to be part technician, part physiologist, and part psychologist to get the job done right.

This multi-view perspective is important because much of the equipment we'll be covering in the following chapters provides essential input to the boat operator. He or she must be able to easily acquire this data, sort it out mentally, absorb it, and analyze it in a timely manner (this is especially true in high-stress situations such as bad weather). It is not my intent to overstate the importance of this subject, but over the years I've seen too much equipment installed that was seriously handicapped in its basic functions because the installer didn't think about the operator's needs when underway. And that is the key point: Installers should position equipment so that it is usable *while the boat is underway.*

So let's put on our physiologist's hat for a minute and see what we need to consider. First, think of the senses we use when operating electronic equipment—sight, hearing, and touch—and ask yourself some basic questions:

- Can the operator easily see the equipment and its screen or display?
- Can the display be read in bright sunlight?
- Can the display be read when the operator wears polarized sunglasses?
- Is the display so bright that it disturbs the operator's night vision?
- Is the display mounted in such a way that it reflects off the wheelhouse window and obscures the view ahead?

Viewing angle is an important consideration for many of these questions, as is the availability of display settings for adjusting brightness and/or selecting "night viewing." While display settings are a function of the equipment, they're also an important component of the overall ergonomic situation. A display that can't be read in the prevailing situation is just as worthless as one that doesn't work.

Similarly, consider hearing and touch. Let's use a VHF radio installation as an example. Is it mounted where the operator can hear incoming transmissions over the sound of the engine? Can the controls be easily reached and accurately adjusted while the boat is underway, even in rough seas? These and similar factors will contribute to or detract from the effectiveness of a given installation.

Electronics installations in open cockpits, such as in most sailboats, can be particularly challenging because direct sunlight can wash out display screens. In some cases, sunlight can make equipment overheat, causing display screens to shut down. While screen technology has come a long way in the last ten years, and the term "daylight viewable" is now a common marketing term used by many equipment vendors, the bottom line is that some equipment is just better suited to a particular location. If you have to leave the helm

Center consoles present an installation challenge as there's not a lot of room for what is often a lot of desirable electronics. Given those limitations, this installation is quite good from an ergonomic point of view. Assuming that the helmsman will be standing most of the time, most of the controls are quite visible and accessible. There is, however, the potential for magnetic interference between the VHF radio and the steering compass, so this may not be the best location for the VHF radio.

electronic compass

chartplotter/
radar display

VHF

autopilot
control

engine monitors
(dual outboards)

Both ergonomics and the potential for magnetic interference have been dealt with rather effectively here. LCD screen displays, as in the chartplotter located to port, have very low magnetic emissions and, although mounted close to the compass, should have no effect. The T-top on this center console boat will provide shade that will help improve screen visibility even on very sunny days.

depth sounder

multi-data display

wind angle indicator

chartplotter

compass

autopilot control

A typical sailboat binnacle-mount system. These work best under a bimini top to ensure screen visibility. Without a bimini, the installer would have to consider the "day bright" capabilities of the equipment.

and run down to a navigation station to check your position, that's bad ergonomics—and possibly bad seamanship.

WATERTIGHT INTEGRITY

Is the equipment waterproof or water resistant? The levels of watertight integrity of equipment are often hidden in obscure specifications that are typically incomprehensible without reference to the standards. And there are several standards that apply to water-resistant enclosures for electronic equipment.

In the Code of Federal Regulations (CFR) Title 46, Part 110, the U.S. Coast Guard defines *waterproof* as: "Waterproof means watertight; except that, moisture within or leakage into the enclosure is allowed if it does not interfere with the operation of the equipment enclosed. In the case of a generator or motor enclosure, waterproof means watertight; except that, leakage around the shaft may occur if the leakage is prevented from entering the oil reservoir and the enclosure provides for automatic drainage." The original text of this regulation included the following definition: "Waterproof machine means a totally enclosed machine so constructed that a stream of water from a hose with a nozzle 1 inch in diameter that delivers at least 65 gallons per minute can be played on the machine from any direction from a distance of about 10 feet for a period of not less than 5 minutes without leakage."

Another standard, which comes from a resolution made by the International Maritime Organization (IMO), applies to handheld VHF radios used in life rafts. The revised "recommendation on performance standards for survival craft portable two-way VHF radio telephone apparatus" describes several performance issues related to waterproof integrity: "The equipment should: be watertight to a depth of 1 m for at least 5 minutes, maintain water tightness when subjected to a thermal shock of 45°C

under conditions of immersion, and not be unduly affected by seawater, or oil, or both."

Perhaps the most recognized and widely used rating system is known as the IP (ingress protection) rating, which is the work of the International Electrotechnical Commission (IEC). It includes a foreign-object rating as well as a waterproof rating, which runs from 1 to 8, as shown in Table 13-1 on page 134.

An example of how this rating would look is "IP X7." Using Table 13-1, we can decode this rating as follows: X indicates no rating for foreign objects, and 7 indicates protection against the effects of immersion to 1 meter in depth for up to 30 minutes. As another example, IP 67 indicates a piece of equipment is dust tight (6) and protected against immersion (7).

One other rating system you are likely to encounter is the Japanese Industrial Standard (JIS). This standard is nearly identical to the IP standard. Like the IP system, the waterproof grade may range from 1 to 8. The higher the number, the better the protection; for example, a product that is waterproof rated to JIS 6 has been subjected to a test that includes exposure to powerful water jets from all directions with no ill effects. A JIS 7 rated product has been tested to withstand being dropped in the water (up to 1 meter deep) and retrieved fairly quickly. This is a useful rating for handheld VHF radios.

Check the product specification sheet for any or all of these ratings. They can help you determine the appropriate location for a piece of equipment—in an open cockpit versus inside a cabin or other enclosed space on a boat. It's valuable information that can dramatically affect the usable life of the electrical or electronic gear used on boats.

THERMAL LIMITS

In Chapter 5, we discussed the use of an infrared heat-sensing gun to check ambient and equipment temperatures for problems or

| TABLE 13-1 | Environmental Ratings for Enclosures Based on Ingress Protection (IP) Code Designations ||||
1st Digit	Protection Against Foreign Objects	2nd Digit	Protection Against Moisture
0	Not protected	0	Not protected
1	Protected against objects greater than 50 mm	1	Protected against dripping water
2	Protected against objects greater than 12 mm	2	Protected against dripping water when tilted up to 15°
3	Protected against objects greater than 2.5 mm	3	Protected against spraying water
4	Protected against objects greater than 1.0 mm	4	Protected against splashing water
5	Dust protected	5	Protected against water jets
6	Dust tight	6	Protected against heavy seas
—		7	Protection against the effects of temporary immersion up to 1 m
—		8	Protection against prolonged submersion beyond 1 m

Note: The letter X is used to designate no rating in either category.
Source: Underwriters Laboratories

potential problems. Here we'll go over both the NMEA and ABYC standards related to temperature control.

The NMEA standard refers to "equipment that requires air circulation for cooling," which I consider to be any piece of gear that is equipped with a built-in cooling fan, mounted on a heat sink, or has finned radiating plates as part of its enclosure. The NMEA standard states that "sufficient means shall be provided to supply circulation of air behind the equipment displays." And for devices that use cooling fans, it states "there shall be sufficient air circulation to cool the equipment during direct sunlight operation with ambient temperatures of 100°F."

The ABYC is quite specific about temperature parameters in two of its standards. Standard A-31 requires that battery chargers and inverters be designed to operate at an ambient temperature of 122° (50°C) and be able to withstand an ambient temperature of 158°F (70°C). ABYC A-28 requires galvanic isolators that will be installed in machinery spaces be rated to 122°F, while those that will not be used in machinery spaces be rated to 86°F (30°C).

In addition, manufacturers establish their own installation guidelines regarding temperature. ICOM, for example, states in the documentation for one of its SSB transceivers that it should not be mounted in locations subject

The back of this heavy-duty DC-to-AC inverter has air-circulation ports for a built-in cooling fan. If you mount this inverter in an area with high ambient temperatures (as in an engine room), and the fan has only hot air to circulate, cooling efficiency will be reduced, and eventually the inverter will stop working.

heat sink fins   diodes (embedded in epoxy
                        "potting" material, which
                        helps transfer heat to
                        the cooling fins)

The heat sink fins on this diode-type battery isolator must have cool air circulating around them or the diodes will almost surely overheat and fail.

to temperatures below –4°F (–20°C) or above 140°F (60°C).

The bottom line regarding temperature is simple: There may be specific industry standards that dictate how you as a technician or boatowner should proceed, and there may be manufacturer recommendations that apply. In all cases, you should follow the manufacturer's specific recommendations. In the absence of those, look to the industry standards for guidance. You must find the applicable limits and consider how they can be met on your boat or your customer's boat. Sometimes this is as easy as finding a new location for the device in question. Other times, it may require the addition of a cooling fan to facilitate proper air circulation or the installation of louvered vents to ventilate otherwise dead-air spaces on board (see photo page 136).

You can use a miniature "pot fan" to supply forced-air circulation behind enclosed electronics panelboard arrangements. Coupled with some additional louvers to let fresh air into the space, small fans can help tip the temperature scale in the right direction. A word of caution, though; these fans are not typically rated for ignition protection, and therefore they should not be used in places such as battery boxes or gasoline engine rooms where ignition protection is mandatory.

POWER SUPPLY

The scene is classic. The boatowner boards his boat and fires up the radar, color chartplotter, fishfinder, VHF, and whatever other electronics are on board, reveling in the glory of all of that expensive techno-wizardry. Life is good. Then he starts the engines . . . only to see all the electronic equipment blink out.

What's happened here is simple: engine cranking draws a lot of amperage, and in this case it pulled the system voltage below the minimum operating parameters set by the electronic equipment vendors. Fortunately, no permanent harm has been done. Once the engines are running (and assuming the engine alternators are functioning normally), voltage will stabilize, and

the equipment will come back online and function normally. But that momentary loss of power can cause any microprocessor-controlled memory function within a given device to "dump" the memory, which is a major hassle if the unit in question has a lot of data entered. This scenario is not only inconvenient, it's also one of the most common electrical/electronic problems experienced on boats, and it drives some owners crazy.

You can adopt one of three strategies to address this problem:

1 The NMEA recommends installing a separate, isolated battery bank to supply all the electronic equipment. This bank can be fed by the engine's alternator for

recharging, but does not add to the engine starting circuitry.

2 The more common strategy is to have all the batteries supplying all the boat's DC loads, including the electronics package and the engine starter circuit. However, this setup requires that all cable runs be sized to maintain an absolute minimum voltage drop, and that the battery bank have sufficient capacity to crank the engines while maintaining a cranking voltage of greater than 10.5 V.

3 The third option is to install a DC-powered UPS device for the electronics main power supply, such as the StartGuard DC power conditioner from Newmar. These boxes, which sell for about $200, are similar in purpose to

the UPS devices common in computer installations; they contain a small battery that "kicks in" to maintain stable voltage output when there is a drop in input voltage. The primary difference is the power source, which is DC, versus the AC used by land-based computers. The StartGuard UPS has a maximum output of 20 amps. If you need additional current capacity, you can install multiple units.

I recommend either option 1 or option 3. If you are interested in option 1—an isolated and dedicated battery supply for powering electronic equipment—keep in mind that space limitations and added cost may rule it out on your boat. In that case, try option 3, a DC-powered UPS device.

Typical integration of a DC-powered UPS device with onboard electronics.

DC-POWERED UPS DEVICES

StartGuard NS-12-20, Newmar, www.newmarpower.com
C-Power, Charles Industries, www.charlesindustries.com

GUIDELINES SUMMARY

Before we move on to more specific installation concerns in the following chapters, here's a summary of some general electronic equipment installation guidelines:

- Be sure voltage and amperage can be delivered to the equipment in the needed amounts under all conditions, including during engine cranking and activation of high-demand loads like electric winches, anchor windlasses, and bow thrusters. Achieving this may require the installation of a UPS device for the system or extra batteries.

- Voltage parameters of the gear in question must be confirmed by the installer or troubleshooter. Some devices (such as radar) have broad acceptable ranges for operating voltage, whereas others (like SSBs and fishfinders) tend to have comparatively narrow operating ranges.

- Reverse polarity can easily and instantly damage electronics. Beware of reversing the positive and negative feed wires.

- Consider possible spray conditions in the intended installation location, and ensure the equipment matches the situation (check the equipment's IP or JIS waterproof rating).

- Consider ergonomics when mounting gear: will it be visible, audible, and reachable and convenient?

- Be concerned about both RFI and EMI, and provide appropriate levels of separation and shielding for equipment and cable runs throughout the boat.

- Make sure that antennas and transducers are oriented correctly from a directional perspective so that their beam transmission is not obstructed by boat structures.

Installing Communications and Position-Finding Systems and Radar

bus bar

Proper installation of communications equipment, position-finding GPS-based systems, and radar is serious business. Boatowners rely on this gear to ensure their safety and, if necessary, call for help, perhaps in a life-threatening situation. Such equipment is vital to the safe operation of a boat, in contrast to, for example, installing a CD player. Yes, a CD player is a piece of electronic equipment, but if the boatowner can't hear the latest Alan Jackson CD while underway, it's not a crisis. So, please, as you read through the material that follows, approach your work as though you might be helping to save someone's life. Don't cut corners.

COMMUNICATIONS SYSTEMS

Communications systems on boats used to entail perhaps just a VHF radio. But nowadays most boaters consider SSBs, cell phones, satellite TV and/or Internet access basic requirements. And of course, all this additional equipment means more work for marine technicians!

VHF Radio

VHF radio is the most common of all marine electronic equipment used for boat-to-boat and boat-to-land communication. It operates in the frequency range of 156 MHz to 174 MHz. In recent years, features such as digital selective calling (DSC), global positioning system (GPS), and the Global Maritime Distress and Safety System (GMDSS) have expanded the use and functionality of VHF radios by incorporating the ability to transmit time, position, and ownership information in the event of an emergency. While improving safety at sea, these features have altered the VHF's traditional installation procedure, so we'll review the installation process.

Range

In the United States, the transmitting power for marine VHF is limited by the FCC to a maximum of 25 watts. Canadian regulations mirror the U.S. regulations in this area. (Land-based units can have much higher power under the regulations.) I have found that

| | TABLE 14-1 | Line-of-Sight Distance, $D_1 + D_2$, for VHF Radios[1] | | | | | | | | |

| | | | Transmitter Height (ft.) | | | | | | | |
Receiver Height (ft.)	5	10	20	40	60	80	100	200	400
5	5	6	8	10	12	13	14	19	26
10	6	7	9	11	13	14	15	20	27
20	8	9	10	13	14	16	17	22	29
40	10	11	13	15	17	18	19	24	31
60	12	13	14	17	18	20	21	26	33
80	13	14	16	18	20	21	22	27	34
100	14	15	17	19	21	22	23	28	35

1. The distances in the table are based on the formulas $D = 1.17\sqrt{H}$ and $D_1 + D_2 = 1.17(\sqrt{H_1} + \sqrt{H_2})$ where D = distance in nautical miles, H_1 = height in feet of the receiver, and H_2 = height in feet of the transmitter. Reprinted with permission from *Boatowner's Illustrated Electrical Handbook*, second edition, by Charlie Wing

estimating 1 statute mile per watt of transmitting power is a fairly reliable rule of thumb for the expected range of a VHF radio. For example, a 5-watt handheld VHF would have a 5-mile range, and so forth.

With transmitting power limited by law, one of the only practical methods of increasing a VHF radio's effective range is to locate the antenna higher. Since VHF operates on a line-of-sight basis, raising the antenna extends its range by extending the effective horizon (see Table 14-1).

We will look further at issues relating to the antenna and the coaxial cable that connects it to the radio in Chapter 16.

Wiring

Since low voltage to the radio will dramatically affect its performance, proper wiring using the correct gauge to ensure minimum voltage drop, and high-quality termination points, are critical to the installation.

Installation guidelines:

- Use the wire gauge as specified by the manufacturer for the DC power feed and negative return.
- Install wiring to keep voltage drop to the maximum of 3% based on ABYC standards.

Line-of-sight transmit/receive distances for VHF radio. (Reprinted with permission from Boatowner's Illustrated Electrical Handbook, second edition, by Charlie Wing)

The wiring and cable connections to a VHF radio. Notice the interface with the GPS receiver via an NMEA cable link.

- Follow Standard E-11 to determine the appropriate wire size to use. Remember that wire size is based on the length of the wire run from the power source to the VHF radio (see the ABYC voltage drop table, page 43).

DSC and GMDSS

DSC is integrated into all new VHF radios and is a part of GMDSS. The interface between the VHF radio and GPS receiver allows position information to be transmitted in the event of a DSC emergency call; all you have to do is push the emergency button now found on all new radio sets to transmit your boat's exact position and time data to the U.S. Coast Guard.

Installation guidelines:

- Use an NMEA-approved cable attachment to link your VHF radio to the GPS

unit (see above illustration). (This interface may be part of your boat's network; see Chapter 17 for more on networking.)

- Follow the manufacturer's recommendations for the exact terminal connections.

As the last step to a fully capable DSC installation, the boatowner must acquire a Maritime Mobile Service Identity (MMSI) number. Once entered, this nine-digit number becomes the permanent identifier for your boat. When DSC is activated, the MMSI will provide rescue personnel with pertinent boat and owner information along with time and position data.

In the United States, you can get an MMSI from the FCC for a fee, or for free by contacting BoatU.S. (www.boatus.com/mmsi) or Sea Tow (www.seasmartvhf.com). In Canada, all applications for MMSI numbers must be

DEVIATION THE HARD WAY

A few years ago, I learned my lesson about electronics-induced compass deviation the hard way—almost literally. My 25-foot walkaround has a very compact helm station, and all the electronic equipment—GPS chartplotter, fishfinder, water temperature gauge, and VHF radio—was tightly clustered at the helm. In the center of the cluster, and no more than 12 to 18 inches from any of them, was the magnetic steering compass.

When my VHF radio failed, I bought a new unit, complete with all the bells and whistles, and mounted it in the same location as the original unit. I tested it and it worked fine. Several weeks later, I was fishing on Narragansett Bay when a thick fog rolled in. I decided to head for home, and checked the chartplotter for the magnetic heading I would need to get to the breakwater at the entrance to my home port.

Off I went, paying more attention to the compass than the chartplotter and following the predetermined heading using the steering compass. About the time I thought I should see the breakwater, it suddenly appeared about 100 feet dead ahead—not the *opening* in the breakwater, where my heading should have positioned me, but the *wall itself*. Fortunately I was cruising slowly at 5 knots, and was able to avoid disaster. But what had happened?

After replotting everything, I realized that I had a 12-degree deviation in my compass. The cause was simple to figure out. I'd never had a lick of deviation in that trusty Ritchie compass until I installed the new VHF. The speaker magnets in the new radio were much more powerful than those in the old unit, and they had caused this new deviation.

I removed the radio and reinstalled it in a new location, away from my compass but still ergonomically placed for ease of use. Another lesson learned, and almost the hard way—breakwater hard!

receiver. The procedure for entering the MMSI is radio specific—it is imperative you have the owner's manual for your radio in hand.

Location, Location, Location

As mentioned in the ergonomics discussion in Chapter 13, the mounting location for the radio is a concern, not only from the point of view of ergonomics, but also regarding EMI. A VHF has two built-in magnets, one for the cabinet speaker and one for the microphone. If the radio and compass are too close together, stray magnetism from the VHF will interfere with the boat's steering compass. The NMEA recommends that the radio and its microphone should be no closer than 3 feet from the compass or any onboard antenna.

Note: Always check for compass deviation when installing electronic equipment of any type near a magnetic steering compass, regardless of whether the equipment has built-in magnets or not. Look for compass needle movement as you place or activate electronic equipment nearby, and modify your placement as needed.

Single-Sideband Radio

Single-sideband (SSB) radio is the long-range counterpart to VHF radio. Where marine VHF is limited to line-of-sight communications and a maximum of about 25 miles, SSB radio has the potential to transmit and receive over hundreds or even thousands of miles. An SSB radio bounces radio signals off the lower layers of the earth's atmosphere, which in turn reflect the signals back to earth. (You can think of SSB transmissions as giant triangles spanning many miles.)

Voltage and Current Requirements

An SSB's appetite for amps covers a fairly broad spectrum, ranging from about 2 to 3 amps in Standby/Receive mode to about

submitted to the Canadian government directly via Industry Canada. To learn more about their process, go to http://strategis. ic.gc.ca/epic/ internet/insmt-gst.nsf/en/ sf01032e.html.

Once you acquire an MMSI number, be careful to enter it correctly into the radio

30 amps when transmitting. This is not too much of a problem unless the operator is the chatty type or spends a lot of time downloading weather charts via a modem. However, 30 amps of consumption is a considerable amount, and it will drain batteries rather rapidly if they are not being recharged.

Installation guidelines:

- Be sure to use correctly sized power feed wires to serve the unit, especially in consideration of restricting voltage drop to a maximum of 3%. Based on ABYC E-11, you will need a minimum of 6 AWG (13 mm²) wire to supply power to a 12 V SSB set over a 20-foot cable run.

- Adhere to voltage supply parameters. Most if not all SSB units operate at a nominal 12 V. For a boat with a 24 V or 32 V system, you will need a DC-to-DC power converter. Overvoltage supply to the transceiver will damage the unit quicker than you can snap your fingers! A typical 12 V (nominal) SSB has a voltage operating range of 13.6 VDC, ±15% maximum. Consider this 15% an *absolute* maximum because performance will definitely be affected if power falls below −15%.

- Use the 3% voltage drop parameter to ensure peak performance. Lower-than-acceptable voltage will have a profound impact on power output from the unit and its potential range. It may be useful to think of potential power out of the unit as proportional to the power available to operate the unit.

- An SSB requires a first-class grounding system to function at its peak, and the antenna needs what is known as a *counterpoise*, a ground that is an integral part of the antenna system (discussed further in Chapter 16).

- For best results, use the multicable harness provided by the SSB manufacturer (see next page), which typically includes the coaxial antenna lead, power cables, and a cable to connect the automatic antenna tuner (a commonly specified available option). All the individual cables in the harness are terminated with watertight connectors for the tuner or transceiver, and all the individual conductors within each cable are connected to the right pins. Stick with the original manufacturer's harness! This is not an area where you will win high marks for creativity.

Regulations for SSB use vary from one country to another. In the United States, FCC rules apply. Additionally, specific regulations apply if the radio includes DSC and GMDSS.

This is where a true electronics equipment specialty house can put you on the right course. Electronics specialists are usually FCC licensed and will be well versed in the specific regulations as they apply not only to installation of equipment, but also to licensing requirements for equipment operators. Inquire before you get in trouble, because the rules changed a few years ago. In some cases, an FCC license is required to perform installations.

Noise Issues with SSB

Electrical "noise" on board can dramatically affect an SSB's ability to produce clear, static-free signals in Receive mode. Potential noise emitters include anything that relies on an electrical spark for its operation—virtually any motor, fluorescent lights, and relays and solenoids. After installation, test your SSB to ensure that the antenna is not receiving any emitted noise.

Since SSBs operate over a fairly broad frequency spectrum (from 2 MHz to 22 MHz), it is hard to predict which electrical devices might contribute to excessive background

A typical cabling configuration for an SSB radio. On a high-end boat today, you will probably see a Pactor modem connected as an auxiliary device for sending and receiving e-mail via radio. The boat's laptop PC will also be linked to the SSB to provide a display and input device for e-mail, weather charts, and the like. Specific wire terminal identification will vary from one system to another, but this illustration shows a general layout. The connecting cables shown are all multicable harnesses with the exception of the antenna coaxial and the wired connection between the antenna tuner and the counterpoise.

noise. To help you find the culprit, conduct a process-of-elimination test, as follows:

1 Set the radio between channels at the low end of the spectrum (towards the 2 MHz end).

2 Activate other AC and DC equipment one at a time.

3 Run all the engines, including AC generators.

4 Repeat this test sequence at all of the commonly used frequencies to be sure that you haven't missed an annoying RFI emitter running at a higher frequency.

Keep in mind that some background noise is normal and unavoidable. But if, for example, audible noise dramatically increases when you turn on a fluorescent light, you've isolated a culprit. Also keep in mind that some electrical equipment is cyclical and may only function intermittently (such as bilge pumps). Check cycling loads when they're actually operating, not just with their power activated.

If you identify a noise source during this elimination process, it's best to consult the radio manufacturer or its local dealer to determine the best type and rating of filter to install.

Your concerns, as the electrical installer, are largely limited to those mentioned: the power supply, cabling, noise, antenna, coaxial cable, and antenna counterpoise. Any problems not related to these issues must be addressed by an electronics technician.

Satellite Communications Systems

Whether they're used for global telephone connectivity, Internet access, or simply tuning into a favorite television or satellite radio show, satellite communications (satcom) systems

Satellite domes are showing up on a lot of boats lately, due to our desire to stay connected.

have gone mainstream on recreational boats over the last decade. The need (or perhaps *perceived* need) of boatowners to stay connected in one way or another has driven this change. As a result, boatowners today have a variety of choices in satcom systems. The capabilities of most of the common two-way satcom systems are shown in Table 14-2 on page 146.

Satcom Installation

Most of the usual installation considerations apply to satellite systems:

- Maintain the usual 3% maximum voltage drop to ensure peak performance, especially since communications systems play such an important role in vessel safety.

- Use appropriately sized wire. Current draw is fairly low, ranging from about 2.5 amps to about 5 amps, so wiring is not as much of a concern as with, for example, an SSB.

- Do not alter the length of the antenna cable (as with almost all coaxial cable installations) or you may seriously degrade performance.

A typical satellite-based telephone/data system using the Inmarsat Mini-M satellite system.

TABLE 14-2

Common Satellite Communications Systems

Application	Inmarsat Fleet 55 and 77	Inmarsat Fleet 33	Inmarsat Mini-M	Inmarsat-C	Inmarsat Mini-C (eTrac)	Iridium	Globalstar	SkyMate
Voice	yes	yes	yes	no	no	yes	yes	no
E-mail	yes	yes	yes	yes	yes	yes	yes	yes
Internet and attachments	yes	yes	yes	no	no	yes	yes	no
Fax machine capable	yes	yes	yes	no	no	no	yes (with additional equipment)	no (outbound only)
Automatic weather/safety	no	no	no	yes	no	no	no	no
Receive weather charts	yes (via Internet)	yes (via Internet)	yes (via Internet)	text weather	no	yes (via Internet)	yes (via Internet)	yes (radar plus selected charts)
GMDSS/Distress	yes (77 only)	no	no	yes	no	no	no	no

SATELLITE TELEVISION AND RADIO

The systems listed in Table 14-2 are strictly satellite *communications* options. DirecTV, a popular land-based satellite company, is one of the most common service providers for boat-based systems as well. Manufacturers of satellite television equipment include:

King Controls, www.kingcontrols.com

KVH, www.kvh.com

Raymarine, www.raymarine.com

Sea Tel, www.seatel.com

Shakespeare, www.shakespeare-marine.com

The two largest satellite radio service providers in North America are:

Sirius, www.sirius.com

XM Radio, www.xmradio.com

- Closely follow the equipment manufacturer's recommendations for antenna placement and separation from other types of antennas mounted on the boat.
- When planning electronics installations, keep in mind possible interference sources in either direction.

You can use the above list to confirm that the electronics specialist has installed your system(s) correctly. Satcom systems are vulnerable to both radio frequency and magnetic interference. Satellite television receivers operate in the 1.5 GHz to 1.66 GHz frequency spectrum, and they can cause interference with cellular telephones (either 850 MHz or 1.9 GHz) and GPS (1.5 GHz). However, interference goes both ways: the GPS or cell phone can also interfere with television reception.

Other satellite voice and data systems, such as the Inmarsat Mini-M system, also operate in the 1.5 GHz to 1.66 GHz range, and are equally susceptible to cross interference. In both cases radar scanners can cause interference, depending on the specific operating frequency.

All satcom receivers use fluxgate, or electronic, compass sensors to help the antenna track and aim toward the appropriate satellite. In some units this sensor is mounted under the same dome as the antenna, while in others it is mounted elsewhere, typically under settees, behind panels, etc. No matter where the sensor is mounted, however, make sure it is far enough away from sources of magnetic interference, including devices with permanent magnets or electromagnets, high-current-draw DC conductors, and large ferrous objects. Remember that compass sensors (and GPS antennas) must be mounted as close to both the port-starboard and fore-aft centerlines of the boat as possible to minimize the effects of the boat's pitch and roll and to provide the sensor with as much stability as possible.

Finally, the NMEA electronic equipment installation standard recommends that a placard be placed near this satcom sensor that reads: "Magnetic Compass Area! Do Not Install Or Store Magnetic Material Within 3 Ft." That recommendation nicely sums up the concerns with magnetism.

Troubleshooting Satellite Systems

A proper satcom installation typically concludes with a series of calibration tests for the fluxgate sensor and operational tests for the receiver. These tests are for *electronics* specialists only, however, and are therefore beyond the scope of this book.

However, as the owner or electrical technician, your troubleshooting should confirm the basics of a proper *electrical* installation as follows:

- wiring links between the system's components are in order
- power is getting to the devices
- any manufacturer-required grounding is in good order
- sources of RFI and EMI have been avoided or eliminated

There are many systems available worldwide, as shown in Table 14-2. The onus is on the electronics technician to be aware of the special needs of this equipment. Aside from minor troubleshooting of the sort I've described above, most issues related to satcom equipment are best dealt with by factory-trained specialists. In fact, many warranties require that the equipment only be installed by such specialists.

POSITION-FINDING SYSTEMS

Over the course of just a few decades, navigation has evolved from paper charts, dead reckoning, and sextants to almost universal penetration of electronic methods. It began with systems like Loran-A and Loran-C, and early satellite navigation (satnav), which gave a position fix every 90 minutes or so (if you were lucky). Then electronic position finding evolved to become virtually synonymous with the satellite-based global positioning system (GPS), which gives us essentially real-time information of phenomenal resolution.

The first GPS units became available to boaters in the late 1980s. They held a lot of promise, but were extremely expensive, and the fact that the full satellite system was not in place left voids in the coverage at certain times of the day in some geographic locations. At one point in the early 1990s, the system seemed to be working quite accurately. Then the Gulf War began, and the U.S. government imposed Selective Availability (SA), which "detuned" the inherent accuracy of the GPS system for military reasons. During this time, however, enhancements were introduced to recover the inherent potential accuracy of GPS. These included Differential GPS (DGPS), which was a supplemental shore-based radio system similar to Loran-C, and WAAS (wide area augmentation system), a system of additional satellites and ground stations that corrected GPS signals, further improving the system's accuracy. SA was turned off by President Clinton in 2000, but it could return at any time.

As of this writing, the industry seems to have settled on WAAS-enabled equipment, which is accurate enough to pinpoint your location anywhere on earth to within 3 meters (about 10 feet) 95% of the time.

Today, GPS is the heart of virtually every electronic navigation system and is perhaps one of the most interfaced devices on board. To use NMEA parlance, GPS is the primary "talker" aboard most modern boats, sending its data to many other devices—the "listeners." It's quite common to have a single GPS receiver linked to one or more repeater displays, an autopilot, a VHF radio, radar, and a computer. Proper GPS installation and performance is key to both the functionality of many devices on board and the safety of the boat and crew.

GPS Installation and Troubleshooting

The major GPS installation component is its antenna, so we'll cover that here, instead of in

GPS antenna

A cockpit coaming is a really poor mounting location for a GPS antenna—right where crew can sit on it and block satellite reception. Not only is it vulnerable to being stepped on and kicked, it may also snag jibsheets, and it even interferes with the coaming's otherwise comfy function as a seat or armrest.

Chapter 16 where other antennas are discussed. Compared to many other electronic systems on board, GPS is *generally* not sensitive to outside interference (although there are some specific instances discussed below). This, at least, makes installation relatively easy.

A GPS receiver is a line-of-sight device. But unlike VHF radios where the line is horizontal, the GPS line of sight is vertical. Its small, typically hockey puck–shaped antenna must have a clear view of the sky to receive signals from the GPS satellites. Anything blocking the sky view also blocks signals, so careful attention must be given to the location of the antenna "puck."

Think about ergonomics here as well. I've recently noticed an alarming trend of mounting GPS antennas on the cockpit coamings of sailboats. Recently, I was on board a test boat with just such an antenna mounting. While beating down Chesapeake Bay, I noticed that the GPS unit was no longer displaying position or speed over ground. I looked around the cockpit and, sure enough, a crewmember

was sitting on the little hockey puck, blocking its reception of satellite signals!

Installation guidelines:

- Match the brand and model number of your GPS antenna with the receiver and the cable harness connecting them.

- Position the GPS antenna as follows:

    out of the way of people and normal on-deck traffic areas

    with a clear view of the sky above

    as close as possible to the boat's centerline

    not too far aloft; the boat's rolling and pitching motion, amplified by increased height abovedeck, will tend to send the GPS a bit crazy trying to figure out where it is

- Because the cable is a proprietary one, do not alter its length.

- Flake excess cable in large elliptically shaped loops. Do not coil tightly. (Vendors typically identify minimum radius bends for cabling in their instruction manuals.)

- Secure flaked cable a sufficient distance (usually 18 to 36 inches) from potential sources of electrical interference.

The matter of flaking the cable versus coiling is one of the most common problems seen in the field. Coiling these cables creates an undesirable ancillary antenna—i.e., separate from the actual antenna at the end of the cable—that easily receives interference.

If the GPS has an additional differential antenna hookup onboard, interference can become a real concern. DGPS operates in the 150 kHz to 500 kHz spectrum and is susceptible to the noise generated by the various sources of RFI already discussed. Appropriate separation (again, typically 18 to 36 inches) needs to be maintained between these emitters and the differential antenna.

In contrast, the operating frequency of nondifferential, or WAAS-enabled, GPS is 1.5 GHz. Most small boats don't have anything on board that approaches that frequency, except for cell phones. As noted earlier, cell phones operate at either 850 MHz or 1.9 GHz. Basic GPS operates roughly in the middle of that spectrum, so a cell phone interfering with a GPS is a real possibility.

Although still rare on anything but the largest boats, satellite television is becoming increasingly popular, and its 1.5 GHz to 1.66 GHz frequency spectrum is close enough that it could also interfere with GPS.

Volts, Amps, and Temperature

GPS is much more forgiving of variations in the DC power supply than some of the other equipment we'll discuss. The voltage operating range is usually from about 10 V to 16 V (although you should always confirm this with the manufacturer's installation documentation). Current draw typically ranges from milliamps

for the smaller units to as much as 4 amps on some of the larger, color chartplotter units with bright backlighting. Therefore, wiring for power to GPS systems does not impose any unusual demands. Some units have a chassis ground lug, and others don't. Use it if it's there, and don't worry about it if it isn't.

Likewise, cooling isn't much of a problem on most smaller units, as they do not generate much heat. On some of the larger displays, the LCD backlighting can generate a fair amount of heat buildup. I use a simple rule of thumb: If there are

GPS can be used on board with a computer. The computer serves the chartplotting function with higher power than a typical integrated chartplotter. The GPS unit can be a stand-alone fixed sensor without its own display, or a GPS chartplotter. The computer accepts data from a variety of sensors in this configuration through a multiplexer (MUX). The multiplexer sequentially forwards the data to the computer from each source. Wind, depth, water temperature, radar, and GPS data can all be presented on the computer screen using navigation and other software. The computer then can direct information and commands to the autopilot. Note that power and ground lines are omitted in the diagram for simplicity. (Reprinted with permission from GPS for Mariners by Bob Sweet)

cooling fins on the back of the display, heat is a concern, so as an installer I need to think about getting air circulation around those fins.

Interfacing GPS with Other Equipment

Interfacing GPS with other equipment, such as autopilots and repeater displays, can be one of the biggest challenges in working with GPS, although this is generally less of a problem if the other units are supplied by the same manufacturer. The GPS unit calculates a lot of data as your boat moves through the water, then it updates the data in real time, and sends it over a network to other pieces of electronic equipment, such as a DSC-equipped VHF, autopilot, and radar display (see illustration opposite). We'll look further at networking in Chapter 17, but specifically as it applies to GPS, there are several things to point out here.

First is the network protocol. Much of this has to do with the digital language the various devices in the network are using. The two most common networking protocols for GPS are NMEA 0183 and NMEA 2000 (see also Chapter 17). There are several versions of NMEA 0183, however, and even within these standards, the "sentence structure" of the digital messages may differ from one manufacturer to another. Some manufacturers don't use NMEA at all, but go with proprietary networking schemes based on Ethernet, a common office network, or on controller area network (CAN), a common automotive architecture.

The bottom line is that it can be difficult to achieve complete compatibility between devices from different suppliers. Some data may transfer readily through the network and some may not.

The only way to ensure compatibility is to locate the equipment's network version number and specific sentence structures in the specification manual for any given piece of equipment. Then compare this information to specifications for the piece of equipment you are attempting to link via the network. Obviously, if you do this *before* buying a new piece of equipment, you'll have the luxury of working with equipment that's designed to work together.

Some examples of the specific codes found in the sentences that talkers and listeners must be able to transmit and receive are shown in the sidebar.

It may be possible, however, to network devices that utilize different protocols or languages with ancillary "black boxes" that are installed between talkers and listeners. Basically these boxes are translators, whose purpose is to

---

COMMONLY USED NMEA 0183 SENTENCE CODES

APB = autopilot, format B
BOD = bearing, origin to destination waypoint
BWC = bearing and distance to waypoint (great circle)
BWR = bearing and distance to waypoint (rhumb line)
DBT = depth below transducer
GGA = GPS fix data
GLL = position in latitude and longitude
GSA = GPS DOP (dilution of position) and active satellites
GSV = satellites in view
HDM = magnetic bearing
HSC = command heading to steer
MTW = water temperature in degrees Celsius
RMB = recommended minimum navigation information when a waypoint is active
RMC = recommended minimum specific GPS/transit data
RTE = route information in active route
VHF = water speed and heading
VTG = track made good and ground speed
VWR = relative wind direction and speed
WCV = waypoint closure velocity
WDC = distance to waypoint
WDR = waypoint distance (rhumb line)
WPL = waypoint location
XTE = cross-track error, measured
XTR = cross-track error, dead reckoning

Black boxes like this one are often required to allow talker and listener devices to communicate across different protocols or languages. While not perfect, they meet the need.

take the language the talker is using and translate it for the listener. You may also find an additional black box, called a multiplexer, that will distribute data to multiple instruments. See the sidebar on page 191 in Chapter 17 for more on networking terminology.

How do you use them? One example would be to connect a combiner/multiplexer box to a piece of NMEA 0183 equipment with a computer via a RS-232 port, or connect to a Raymarine device that is networked via that company's proprietary SeaTalk system.

RADAR

Similar to satcom, radar may require the services of a manufacturer-trained specialist to set up the initial calibration and tuning of the unit. This need will depend on the boatowner's familiarity with radar. In many cases, a trained professional is the best choice for an initial run-through of the unit's operational characteristics. Although all these units operate on the same

principles, software access, system menus, and control functions can be confusing at first, and often the instruction manuals provided with these units are absolutely confusing!

This does not, however, preclude boatowners or marine electricians from performing a radar installation if they desire. Most radar units marketed for the recreational boating market are "pre-tuned" and calibrated at the factory, so there are only a few issues to deal with.

Installation guidelines:

- You must use the cabling provided by the equipment manufacturer.

- You must not alter the length of those cables. If you need longer cables than those supplied, all the manufacturers offer optional longer cables, which you must use to ensure proper operation.

- Consider the ergonomic issues already discussed—viewing angle, display settings, etc.

Perhaps the biggest concern with radar is that the highly focused beam of energy in the microwave frequency range it transmits may be harmful to people on board the boat, and also can affect other onboard electronic equipment. Most marine radars units installed on recreational boats are X-band units, which operate in the 8 GHz to 12 GHz frequency spectrum. Some larger commercial craft may also have S-band radar units, which operate in the 2 GHz to 4 GHz spectrum.

The matter of whether or not radar is harmful to people on board the boat is debatable. Cases of testicular cancer, cataracts, and even behavioral changes have been attributed to exposure to radar beams. After analyzing data provided by the World Health Organization (WHO), it seems clear to me that you would have to be literally sitting in front of the radome, say within 3 feet, for extended periods of time to cause any ill effects. And even that is not certain. In the United States, groups such as the National Council on Radiation Protection (NCRP) are researching the possible carcinogenic, reproductive, and neurological effects of human exposure to the levels of radiation emitted by radar units. One thing we do know for sure is that all the radar vendors will discuss the potential safety hazards associated with exposure to the radar beam and make installation suggestions that will minimize the possibility of this occurring.

As far as interference with other electronic equipment is concerned, WHO sums it up like this: "Radars can cause electromagnetic interference in other electronic equipment. The threshold[s] for these effects are often well below guidance levels for human exposure to RF fields. Additionally, radars can also cause interference in certain medical devices, such as cardiac pacemakers and hearing aids. If individuals using such devices work in close proximity to radar systems they should contact [the] manufacturers to determine the susceptibility of their products to RF interference."

Ensuring Electromagnetic Compatibility

Radar manufacturers are concerned about electromagnetic compatibility (EMC), which is the ability of electronic equipment to operate in proximity with other electrical and electronic equipment without suffering from, or causing, impaired performance. For our purposes, it is essentially synonymous with EMI. Improper installation of a radar set will almost certainly guarantee problems with other equipment as well as with its own performance. So we need some commonsense guidelines to help us achieve compatibility:

- Keep radar and its connecting cables at least 3 feet (1 m) from any other transmitting equipment or cables carrying radio signals, such as VHF radios. In the case of SSB radios, double this separating distance because SSBs are more powerful than VHF radios.

- Keep other antennas and electronic equipment out of the radar beam's path by at least 6.5 feet (2 m). The radar beam typically spreads about 12.5 degrees above and below the radiating element (scanner) on the radar.

- Radar needs a power supply separate from that of the engine starting motors. Equip it with its own battery or a UPS device (such as the Newmar unit mentioned in Chapter 13) to prevent the unit from blinking out and having to restart. Voltage spikes and/or drops that occur during engine cranking will often be outside the voltage-operating parameters for the radar.

- If the cabling is equipped with a noise suppression ferrite as shown in the photo,

Two typical power cord ferrites. These small devices are increasingly found on wiring harnesses and power supply cords on modern equipment, and they are crucial in helping to minimize interference effects.

do not remove it when pulling cables through the boat during installation. If the cabling has to be removed, it must be reassembled in exactly the same way and in the same position.

These considerations are universal and should be carried out for all radar system installations.

Power Supply

As with satellite communications systems, radar systems have multiple power-supply requirements: one to the display unit or receiver, one to the antenna or scanner, and possibly one to the scanner's motor drive. Different equipment manufacturers take different approaches; some of them use multiple

A typical radar system layout with peripheral electronics. This is a relatively simple point-to-point arrangement. Radar can also be integrated into an onboard network to share data with several other devices, such as autopilots, repeater displays, and chartplotters.

power sources, and others use a single, split source. In the end, the manufacturer's wiring system diagram is the only practical way to find the appropriate power supply.

Power supply conductors are often bundled into a harness with the signal-carrying conductors. Remember, it is essential to use the manufacturer's cable harness. If you have to alter the cable length, some manufacturers offer harnesses in a variety of lengths to accommodate different installation requirements. Pay close attention to any specified minimum bend radius. These harnesses may not come supplied with fuses to protect them in the event of an electrical overload or short circuit; be sure to use the *manufacturer-recommended* fuses and locate them where specified.

Installing Depth Sounders, Weather Systems, and Autopilots

D epth sounders, weather monitoring systems, and autopilots are three areas in marine electronics where we have seen dramatic improvements and increased sophistication over the last ten years. Depth sounders and particularly fishfinders have evolved from the old Si-Tex paper drum recording units that were a mainstay among serious fishermen for some time to the full-color LCD display units of today. Side-scanning and even forward-searching depth-sounding capabilities are now available for those who feel the need for that level of underwater probing. As for weather monitoring, the old wind vane and cup masthead sensors in use for decades are being replaced with ultrasonic sensors that have no moving parts to wear out or get broken by errant seagulls. Autopilots have gotten more powerful, and draw less electrical current than they used to, enabling offshore cruisers to consider using electric autopilots instead of wind vane controlled self-steering systems (which are always a concern due to battery capacity limitations). They are now relatively easy to interface with GPS, thus not only automatically steering your boat, but taking you where you want to go quite accurately.

DEPTH SOUNDERS

Depth sounders and fishfinders are echo-sounding systems used to determine the depth of water under your boat or to find targets (many units combine the two systems). They are similar to most of the devices discussed in Chapter 14 in that they have a display, transmit and receive signals, and use cables. They are different in that instead of an antenna, depth sounders have a transducer, which for most recreational systems transmits at a frequency of 50 kHz or 200 kHz. (Lowrance—www.lowrance.com—recently began advertising dual-frequency units transmitting at 80 kHz or 200 kHz.)

Depth sounders can be simple devices that only tell you the distance to the bottom or they can be feature rich with, for example, various alarms (shallow/deep water, warning), forward scanning, and the ability to provide pictures or a graphic contour of the bottom. Some include SeaTalk or NMEA capabilities and can be networked to your autopilot, GPS, radar, or chartplotter. They range in price from $100 to $3,500.

Depth sounders and fishfinders can be among the most troublesome of all categories of marine electronics because they are sensitive to both over- and undervoltage situations. In addition, the transducer that sends and receives signals from the bottom can be fouled by marine growth and needs periodic cleaning. It also must be carefully installed in the correct location on the boat's bottom to ensure maximum functionality. To keep these devices performing properly, a first-class installation is paramount.

Displays and Cables

Depth sounders have four types of displays: LCD screen (color and black and white), CRT (similar to a television screen), paper graph, and spinning wheel flasher. When choosing a depth sounder or fishfinder, be sure to check the readability of the screen, especially in daylight.

The cable that connects the transducer to the display is carefully impedance matched to the system, so under no circumstances should you alter its length or wire gauge. Doing so would severely affect the performance of the system. When installing, carefully route the cable to avoid interference from other electrical equipment.

Transducers

The transducer is mounted on the bottom of the boat. Its purpose is to convert the electrical signal, or pulse, from the depth sounder to a sound, called a *ping*, and transmit it through the water. The sound wave travels until it hits something, and then echoes back. The transducer captures the echo, converts the sound back to an electric signal, and sends this back to the depth sounder. This process of transmitting, echoing, and receiving can happen up to forty times per second.

Transducers on simple, numerical depth sounders are intended to point straight down (or nearly so), while on the most sophisticated units they may have sideways- or forward-transmitting

The beam of a high-frequency (200 kHz) depth sounder is narrow, covering a 10- to 15-degree arc. This is good for close focus on the bottom and will give maximum bottom detail. At 50 kHz, the beam is wider and generally better for scanning greater depths and providing a broad view with fewer details. Another consideration is power output. A 1,000-watt unit can reach greater depths than a 600-watt unit.

paddle wheel

A typical transducer installed on a fairing block on a powerboat. The joint between the transducer and the fairing block is difficult to see because antifouling paint has filled the seam, but the fairing block represents about 50% of the profile surface. This particular transducer also includes a small paddle wheel that drives a knotmeter function on the same instrument.

fairing block

fairing block

Installation of a transducer. Note the fairing blocks both inside and outside to create a level mounting surface and (if this were a through-hull or seacock) spread any accidental load (e.g., being hit by gear or trodden on). Note also the generous amounts of bedding compound. (Reprinted with permission from Nigel Calder's Cruising Handbook by Nigel Calder)

capabilities, enabling you to see what's underwater ahead, or to port and starboard.

As the ping, or sound wave, is transmitted into the water by the transducer, it forms a cone-shaped beam, which can be as narrow as 15 degrees or as wide as about 45 degrees (as shown in the illustration on page 157). Some of the more sophisticated units, such as those available from Interphase Technologies (www.interphase-tech.com), can scan through even wider arcs.

In order to receive the echo, the transducer must be oriented correctly on the hull according to the manufacturer's guidelines. Most often, some sort of fairing block is needed to aim the beam correctly.

Mounting a Transducer

To make the most of your depth sounder/fishfinder, the transducer must be mounted in the best location for your boat (i.e., sail or power). We'll cover general guidelines first, then go into some specifics.

- As much as possible, mount the transducer so it has an unobstructed path for the beam.

- Mount the transducer so it will be submerged at all times while underway. This means powerboat owners must take the location into account when the boat is up on plane.

- Do not mount a transducer in the flow of turbulent water, such as near a prop or behind a strake or through-hull flange. For inboard boats, do not mount it near the propeller shaft struts or raw-water intake fittings.

Transducer mounts on sailboat hulls can be tricky because a long keel is more likely to interfere with the cone-shaped beam as it expands downward from the transducer. Either forward or aft of the keel, rather than to either side of it, is often the best location. Powerboats with prominent keels can also have this problem. If you have a side- and/or forward-scanning unit, it must be mounted on the centerline, forward of the keel.

Transom-mounted transducers are typically used on powerboats due to their relative ease of installation. However, do not place a transducer directly in front of an engine drive's lower unit, where the disturbed water flow could interfere with engine cooling and might cause propeller cavitation as well. It is advisable to offset it by as much as 18 inches. On single-engine installations, this means the transducer will be away from the boat's centerline, while on dual-engine installations, the centerline may be the best location.

Yet another mounting option is to glue the transducer to the inside of the hull. In most cases, the transducer will transmit and receive its signal directly through the hull material; in others, it is mounted inside a fluid-filled chamber (usually filled with compass fluid or a 50-50 blend of antifreeze and water), which is mounted to the inside of the hull. These chambers are available commercially, but they can also be home-built from 4-inch ID (inside diameter) PVC pipe cut to the hull contour and sealed on top with clear Plexiglas and silicone sealer. It used to be easier to replace transducers mounted in this fashion, but with modern adhesives, there's no longer much of an advantage.

For an inside-the-hull installation to be effective, the transducer must rest on a solid mass of material. Solid GRP (glass-reinforced plastic—i.e., fiberglass) and wood hulls work well, but cored laminates do not. Some transducers may work through some metal hulls, but it's advisable to test the transducer first before making the installation permanent.

Whether a transducer is mounted in a chamber or to the hull, the transducer face needs a "solid" surface for its vibrations to be

This transom-mounted transducer on a dual-outboard boat is located to port of the boat's centerline to avoid disturbing the water flow to either of the lower units.

transmitted. You must make an airtight, void-free bond between the transducer face and the hull surface. Use a glob of silicone caulk or 3M 4200 adhesive sealant, being sure to use a sufficient amount of the adhesive to create a solid mass of material between the entire face of the transducer and the hull. *Do not use* 3M 5200 for this purpose, or you will have a very difficult time removing the transducer when you want to replace it.

Installation and Troubleshooting

Now that we've determined the location and method of mounting the transducer, let's move on to installation and troubleshooting.

Cables

As noted earlier, the cable is impedance matched to the system and altering its length will affect system performance. Route cables so as to avoid EMI and sharp bends.

Intermittent problems with depth sounders are often related to cable failures. A "wiggle" test of the entire length of cable will help you identify loose or broken connections under the cable insulation.

To perform a wiggle test (which will work on any long cable runs):

1 Run the system while flexing the cable itself, particularly its termination points.

2 Have a partner watch the display and see if it remains constant.

3 If the display blinks out while you're wiggling at a certain point, you have found a bad connection or break in the wire(s).

In most cases, repair will require replacing the cable with a factory-supplied unit. In other units, the cable may be permanently connected to the transducer, so replacing the entire transducer/cable assembly may be your only option.

Warning: Never disconnect the cable from the display while the system is turned on; it can produce a pronounced electrical arc. Power down the unit before disconnecting the transducer cable.

Power Supply

Depth sounders are quite sensitive to power supply. Most operate in the 10 V to 16 V range and require 1 to 4 amps of current. Symptoms of excessively high or low voltage vary with the complexity of the unit. With a simple numerical read-out sounder, you'll encounter errors in depth readings or a blinking display; with a fishfinder, you'll get garbled and shifting or random bottom contours. The obvious troubleshooting approach is to test and confirm voltage supply to the display unit with your DVOM.

Interference

As with all electronics, you must also be aware of the possibility of interference. In spite of cable shielding, cross-conductor interference often occurs, especially on modern boats in which numerous cables are crowded together in conduits or wire trays. Because the cable connecting the transducer to the display is a fixed length, flake excess cabling in an out-of-the-way location (see Chapter 14). However, if you must coil the cable, be sure to locate the coil far from any of the high-output emitters identified in Chapter 9.

If you suspect interference (after first confirming the quality of the voltage supply and return ground to the system), the only way to isolate the source is through process of elimination: With everything else on the boat turned off, including the engines, check the depth sounder for normal operation. Then, turn on one circuit at a time and see if the problem appears. After you've isolated the culprit, establish an appropriate zone of separation between the offending component or circuit and the transducer cabling, as described in

Chapter 9. You may have to reroute the transducer cable to correct the problem.

Remember that potential interference cuts both ways. Some manufacturers recommend that the display unit be mounted at least 3 feet (1 m) away from any steering compass. This is hard to achieve on smaller boats with tight consoles, such as center console fishing boats. If your boat layout requires a smaller zone of separation, then before you install the unit, confirm that operation of the depth sounder/fishfinder will not cause compass deviation.

Failure

Transducers are subject to failure, but since they are a sound-producing device, you can easily determine if they are functioning by simply listening to them. I typically touch one end of a long wooden dowel or a long screwdriver to the transducer either inside the boat, or if the boat is out of the water, on the face of the transducer itself. For an externally mounted unit on a boat in the water, you can touch one end of a long dowel or boathook to the unit. With the system turned on, put the dowel or screwdriver handle to your ear. If it's working, you should hear a distinct clicking sound.

More empirical results are possible if the manufacturer has provided specific electrical tests to perform at the display end of the cable harness assembly. However, product-specific literature is necessary to identify the right gang plug pin ends and appropriate procedures and results for these tests.

In a nutshell, troubleshooting depth-sounding equipment involves many of the same procedures and limitations as other gear types already discussed, namely:

- Check the voltage quality under all conditions, including during engine cranking.
- Confirm the transducer is properly mounted and oriented.

- Ensure cables and connections are sound and separated by sufficient distances from other electrical devices to rule out EMI.

For problems beyond the scope of these tests, consult an electronics specialist. In most cases, problems with a depth sounder's internals are generally resolved by simply replacing the transducer, display, and/or cable.

WIND- AND WEATHER-MONITORING INSTRUMENTS

In terms of frustration, instruments for monitoring wind direction and speed were near the top of my list for years. This probably reflects my experiences as a navigator on high-performance racing sailboats. Both crew and owners wanted absolutely precise tack-to-tack readings—a goal that was often hard to achieve back then. In the end, if I got it within a few degrees, I felt I had a good day.

The calibration of wind and weather instruments has always been a real chore, and certainly beyond the normal realm of the marine electrician or boatowner. There was nothing simple about this process. Unless you followed the manufacturer's specific calibration instructions, which were often about as clear as reading a manuscript written in hieroglyphics, you didn't get very far.

Most of the problems centered on updrafts—air that flows up a sailboat mast, spilling off both the headsail and the mainsail at the top. An updraft creates turbulent air, which disrupts both the traditional wind vane pointer that provides directional data and the wind cups that provide velocity data. Additionally, these sensors are electromechanical and subject to the usual problems of such devices on a boat. An errant seagull colliding with the vane or wind cup and the failure of a bearing are not uncommon problems.

In recent years, equipment has improved considerably with offerings from companies such as Brookes & Gatehouse (B&G, www.bandg.com) and Airmar (www.airmar.com). For example, Airmar recently introduced a new weather-monitoring sensor/transducer that promises to change the way boaters monitor barometric pressure, wind direction (true and apparent), wind speed (true and apparent), air temperature, dew point, relative humidity, and even wind chill temperature and GPS position data (see below). The sensor is an ultrasonic device, with no moving parts. All functions are displayed on a PC, which also can be interfaced with chartplotters, autopilots, and other devices. Plus it is compatible with NMEA 0183 and NMEA 2000 devices (see Chapter 17), which also permits the use of conventional instrument displays (as long as they are NMEA compatible). They are not cheap, however, costing around $1,000.

WeatherStation Layout

Airmar's WeatherStation system consists of a sensor, a power supply to the sensor, and a converter unit that interfaces with a laptop and is powered by the computer's USB supply. An optional piece of equipment called a combiner lets you connect to a laptop and NMEA device simultaneously.

Installation involves the same considerations as always: pay attention to sensor positioning (see below) and cable routing, and ensure the device has a sufficient power supply. In addition, you will need to install overcurrent protection for the power feed conductors, based on the manufacturer's recommendations.

Sensor placement is critically important; parameters are as follows:

- Proper orientation in the fore-and-aft plane is a must for the unit's directional sensing capability to work.

- To accurately sense airflow velocity and direction, it must be mounted in a location that ensures unobstructed and undisturbed airflow across the sensor surface. Airmar recommends a minimum of 6.5 feet (2 m) of clearance between the sensor and any obstructing object. This will not be an issue with sailboat installations if the sensor is at the top of the mast, but it could be a real challenge on many powerboat installations.

- Placement relative to other transmitting antennas is also important as they can disturb airflow and create turbulence near the sensor.

- The unit is also a GPS sensor, so interference from other antennas is possible (see Chapter 14).

The WeatherStation's total power consumption will depend on its configuration. If the sensor is connected to a PC, then obviously the PC will require a power supply. If NMEA-compatible displays will be used, then each will need its own power supply. The sensor itself operates in the 10 VDC to 16 VDC

Typical configuration of an Airmar WeatherStation installation.

range and requires only 0.5 amp of current, so even for masthead-mounted sensors, the weight of wiring aloft will be low.

As with all wind-machine instruments, especially those with long wire runs up sailboat masts, you must install cables carefully, using proper support and strain relief at the masthead to ensure acceptable service life. While Airmar suggests that cable lengths not be altered, they do provide specific guidelines for making connections and attaching RFI shielding if you have to remove a cable connector to facilitate routing through tight openings.

For the average boatowner or marine electrician, troubleshooting this system will focus on confirming voltage and ground integrity. The TDR tool (Chapter 3) will be useful to confirm wiring harness integrity.

AUTOPILOTS

Autopilot installation and troubleshooting can be critical to the safety of your boat and crew. And depending on your boat, you may need the combined efforts of an electrical installer (possibly you), an electronics specialist, and a marine mechanic.

I say mechanic, because if the boat has hydraulic steering you may need a mechanic to mount and interconnect the hydraulic steering pump that is part of the autopilot system and is integrated electronically to the autopilot control box. Depending on the system you choose, the autopilot pump may replace the existing hydraulic steering pump. In any case, the autopilot pump will become the driving force for the hydraulic ram or actuator at the rudder.

In nonhydraulic steering systems that rely upon gears or cables (or even direct tiller-to-rudderstock connections), a mechanical actuator servo must be linked to the existing steering system. (You may need a mechanic for this also, depending upon your skills and/or those of your electrical technician.)

Autopilots are an interesting mix of electronic, electrical, and electromechanical components that must be integrated with the boat's onboard navigation system to function properly. The system shown is for a boat with hydraulic steering. On a boat with mechanical steering, the pump in the illustration would be replaced with an electric servo and a mechanical connection to the steering system.

The illustration highlights several key points:

- The RF ground serves to minimize the effects of any EMI with the course computer. Equipment vendors' instructions for the connection of this ground must be followed to the letter.

- The data cable interface harness may be split farther away from the course computer to provide and receive data from NMEA-compatible devices such as a GPS, and/or to connect to a proprietary network like Raymarine's SeaTalk or Furuno's NavNet.

- This same harness may also contain cabling to feed a system alarm.

Additionally, it is imperative that the electronics specialist involved with any installation carefully check the interface protocols between the autopilot computer and the GPS receiver to ensure that all the necessary data will be transmitted and received.

Autopilot Installation

Most autopilots will function in a 10 V to 16 V range without incident. The 3% maximum voltage drop guideline applies here as well, so be careful to consider circuit length when calculating power feed wiring requirements. Autopilots are fairly heavy power consumers, drawing anywhere from about 4 amps to as much as 10 or 12 amps, depending on the system. This high current draw is due to the hydraulic pump/motor combination, which is constantly working when the system is on, and motors by their very nature are heavy electrical consumers.

Note: Many of the manufacturer-supplied electric motor drives *are not* rated for ignition protection. These drives *should not* be installed in engine room spaces or fuel tank storage areas on gasoline-fueled boats. Both the U.S. Coast Guard (in CFR Title 33, Part 183, Subchapter I) and ABYC Standard E-11 require that all installations in these areas be rated as ignition protected. Keep in mind that the CFR standards are mandatory; the ABYC standards are considered voluntary.

Unfortunately, I see this requirement overlooked on many boats, and it represents a serious fire hazard if a fuel leak ever develops on those boats. Check with the equipment manufacturer that the motor drive is ignition protected if you plan to install it belowdeck on a gasoline-fueled boat. If it isn't ignition protected, you should swap it out for one that is.

Because autopilots rely on electric motors, proper rating and placement of overcurrent protection for the circuit and/or motor are important considerations for the marine electrician. In fact, this is true for any electrical equipment installed on board, and the best course is usually to follow the manufacturer's recommendations to the letter.

Finally, for the mechanic, it's important that whatever the steering actuation device is—motor-driven lever action or hydraulic pump—it must be securely mounted to the boat's structure. Think through-bolts, not self-tapping screws!

While it's best to leave the final testing and calibration of autopilots in the hands of the electronics specialist, you can provide the same basic quality control functions as with other equipment:

- Make sure the system computer is mounted as far as possible from any RFI or EMI emitters.
- Check for interference by running the boat under autopilot and, one by one, operating all other electrical equipment. Obviously, this test should be done in open water in case interference does occur that causes the boat to suddenly change course.

Coaxial Cable and Antennas

Unlike most other signal and power wiring, coaxial cable (or coax) is not merely a conduit between devices; it represents an inherent part of most marine antennas, and needs to be addressed as a critical component.

COAXIAL CABLE BASICS

We need to focus on coaxial cable selection and installation requirements since coax is the connection between antennas and their transceivers; there are quite a few commonly used variations of coax cable available; and signal losses through it can diminish the performance of a transceiver, or even render it almost useless. After voltage and amperage problems, loss through this vital link is perhaps the biggest factor in poor equipment performance.

Coaxial cable is different from the wire and cable used strictly to conduct electricity or large amounts of electrical current. And although coax is sometimes used to carry electrical power, its more common use, and the one we'll focus on here, is to carry radio, video, and data signals.

Coaxial cable consists of four layers (see illustrations page 166). In the center is a conductor composed of either solid copper wire or stranded copper (commonly used in marine applications). The core conductor is surrounded by a dielectric material or insulator. Surrounding the insulator is the second conductor, a sheath or shield, which can be either braided copper or a foil. The final layer is the outer jacket (insulation), which protects the cable from the environment and provides some flame retardation.

The signal is carried by the center conductor, while the shield protects the signal from RFI and serves as a ground. The design is intended to keep the signal in place, so that it will not "couple" with adjacent wiring routed in the same cable run, and noise from adjacent cables will not interfere with the signal.

Boat-system installers and troubleshooters have three specifications to be concerned with: attenuation, impedance, and velocity of propagation.

Attenuation

Coaxial cable will always have some attenuation, which is the amount of signal loss per given length of the cable. This value is measured in decibels (dB). The idea is to keep the loss within parameters described later in this chapter.

Coaxial cable construction.

Three sizes of coaxial cable with the inner insulation stripped back to show the center conductor, and the outer insulation stripped back to show the (somewhat crumpled) braided shield.

Impedance

Each type of coaxial cable has an inherent impedance, or opposition to current (it's the high-frequency equivalent of resistance). More impedance means more attenuation. Interestingly, this impedance is not related to the length of the cable but to the wire gauge of the conductors and the thickness and effectiveness of the insulation between them.

Coaxial cable specifications typically assign a "characteristic" impedance, which is calculated from the ratio of the inner and outer diameters and the dielectric constant. Assuming the dielectric properties of the material inside the cable do not vary too much over the operating range of the cable, this impedance is frequency independent.

The most common impedance values associated with coaxial cable today are 50 ohm and 75 ohm. Usually, the 50-ohm variety is used to carry radio signals and data (Ethernet), and the 75-ohm variety is used for audio, video, and some telecommunications applications. The attenuation associated with each of these values is related to the frequency of the signals each carries. So attenuation and frequency will help determine the best cable choice and rated impedance for a given application.

Equipment should be matched to the cable's impedance. When radio frequency signals are transmitted through coaxial cabling, the cable impedance is significant in determining the load of the transmission source and the overall efficiency of the transmission. Another way of saying this is that the source impedance must be equal to the load impedance (often the antenna) to achieve maximum power transfer and minimal signal loss. We need to have matched impedance values for the transmission source, the cable, and the receiving device at the other end of the cable. Therefore, make sure the correct impedance cable is used for the application at hand, as recommended by the equipment manufacturer.

Velocity of Propagation

As we covered in Chapter 3, VOP is a percentage of the speed of light that describes the speed of current through a cable. Different conductors have different VOP values due to their inherent resistances to electrical current flow.

However, as coaxial cable ages, its impedance changes, and consequently, so does its VOP value (increased impedance results in lower VOP). This indicates a general degradation of the cable and that replacement is in order. Signs of a poor coaxial connection or degraded cabling will be poor radio reception or low transmission range.

You can check the state of the cable by monitoring its VOP with a TDR (Chapter 3). You can identify cable breakdown or failure, or confirm that the cable is stable and within known specifications.

CHOOSING COAX FOR MARINE APPLICATIONS

In choosing coax cable, we'll look at three specifications: type, attenuation, and impedance.

Type

Coax is identified by various RG types, including RG-8, RG-213, RG-8X, RG-58, RG-59, and RG-6U, as well as LMR, such as LMR-400. These designators can be a bit confusing, but the letters and numbers do have meaning. We'll use RG-8/U as an example:

R stands for radio frequency

G stands for government

8 represents a government approval number

U represents a universal specification

If the letter A, B, or C appears before the slash (for example, RG-8A/U), it indicates a specification modification. The only way to be crystal clear on these subtle specification differences and letter designators is to check out the website for one of the major coaxial suppliers (e.g., Belden, www.belden.com) and compare the specifications closely.

Table 16-1 lists the characteristics of five types of coaxial cable found on boats:

- RG-58/U is very thin and typically used only for interconnecting electronics where lengths are short and signal attenuation is not a problem.
- RG-59/U is used to connect television antennas and cable service.
- RG-8X is typically used to connect VHF and HF antennas up to lengths where attenuation becomes excessive.
- RG-8/U and RG-213 are both used to conduct maximum power to VHF and HF antennas.

Some letter designators don't fall under this set of government rating numbers. A good example is the LMR-400 coaxial mentioned above. In this case, "LMR" is simply a registered

TABLE 16-1 Coaxial Cables

Specification	RG-58/U	RG-59/U	RG-8X	RG-8/U	RG-213
Nominal O.D.	$3/16$"	$1/4$"	$1/4$"	$13/32$"	$13/32$"
Conductor AWG	#20	#23	#16	#13	#13
Impedance, ohms	50	75	50	52	50
Attenuation/100', dB					
@50 MHz	3.3	2.4	2.5	1.3	1.3
@110 MHz	4.9	3.4	3.7	1.9	1.9
@1,000 MHz	21.5	12.0	13.5	8.0	8.0

Reprinted with permission from *Boatowner's Electrical Handbook*, second edition, by Charlie Wing

TABLE 16-2	Maximum Allowable Signal Loss by Equipment Type (per NMEA)	
Equipment	**Maximum Loss (dB)**	**Operating Frequency**
VHF radio	3	162 MHz
SSB radio	3	2–22 MHz
Cell phone	3	850 MHz or 1.9 GHz
Television	6	54–806 MHz
Satellite TV	6	1.6 GHz
GPS	3	1.5 GHz
DGPS	3	150–500 kHz
Source: NMEA		

trademark held by Times Microwave Systems (www.timesmicrowave.com). This cable is generally considered superior to any of the R-designated cables. The company has a lot of experience developing special cable for military applications; LMR cable is one by-product of their research and development.

This discussion may seem cryptic, but the electronic equipment manufacturers generally make the decision for us in terms of what cable to use. For example, Ancor Marine (www.ancorproducts.com) makes some very specific recommendations in its catalog: RG-58C/U for interconnecting electronic equipment; RG-59/U for televisions and some satellite systems; RG-8X for VHF radios in which the distance from the antenna to the transceiver is less than 50 feet; and RG-8/U for VHF with longer cable runs to the antenna.

The NMEA on Coaxial Cable

When all is said and done, however, the best resource for sorting through all of this is the NMEA. It devotes ten pages of its electronics installation standard to coaxial cable requirements—in standards writing, that's significant. (But even the NMEA will tell you to first consider the equipment manufacturer's recommendations.)

The NMEA defines maximum acceptable levels of signal loss for different types of equipment, shown in Table 16-2. Use these values unless the equipment manufacturer specifies a lower value.

Attenuation

Next let's look at the specified attenuation for the different cable types, as shown in Table 16-3. Specifications for attenuation in cable with the same basic designator (various brands of RG-58 or RG-8X, for example) may differ slightly from one manufacturer to another. It is advisable to double-check the values in the table against those for the specific brand of cable you are using, particularly if you're installing long cable runs. It could mean the difference between a pass/fail situation for a given installation.

Impedance

Of course the last spec to match is the inherent impedance of the cable. Some applications call for 50-ohm cable, others 75-ohm cable. Impedance ratings of various cable types are shown in Table 16-4.

TABLE 16-3 — Attenuation in dB per 100 Feet of Coaxial Cable

Cable Size	Frequency (MHz)				
	30	50	100	1,000	2,400
RG-58A/U	2.5	4.1	5.3	24	38.9
RG-59	—	2.4	3.5	12	—
RG-8X	2.0	2.1	3.0	13.5	21.6
RG-8/U	—	1.3	2.2	9.0	—
RG-213	1.2	1.6	1.9	8.0	13.7
LMR-240	1.3	1.7	2.5	8.0	12.7
LMR-400	0.7	0.9	1.3	4.1	6.6

TABLE 16-4 — Inherent Impedance by Coaxial Type

Cable Size	Impedance (ohms)
RG-59	75
RG-58	50
RG-8X	50
RG-8/U	52
RG-213	50
LMR-240	50
LMR-400	50

Other 75-ohm cable sizes sometimes used in marine applications are RG-6 and RG-11, but these are rather difficult to find in most chandleries.

INSTALLING COAXIAL CABLE

When installing cable, be careful about mismatches; that is, connecting two cables together that have different resistance values or connecting cable with the wrong impedance value to a piece of equipment. The longer the cable run, the more this mismatch will affect the performance of the equipment.

Cable Lengths

The NMEA suggests that cable lengths be modified for the specific installation. The coils or bundles of coaxial cable that we so often find hidden behind panels are not desirable or wise. Cut the cable to the length you need and install a matched cable end. Allow a minimum 12-inch service loop at the equipment end. (This, of course, would only apply if the equipment manufacturer does not specifically prohibit the alteration of the coaxial length, which, as discussed in Chapter 14, is often the case.) The goal is to make the most direct connection possible in order to reduce the attenuation between the antenna and the device to a minimum.

In routing coaxial cable, pay attention to its minimum bend radius. Sharp bends in coaxial cable increase attenuation and contribute to signal loss in the run. Minimum radius bends for different cable types are specified by the NMEA, but you should always confirm the cable manufacturer's minimum bend specification for a type of cable, as the numbers vary from one manufacturer to another. Table 16-5 shows minimum bend radii per NMEA requirements.

TABLE 16-5	Minimum Bend Radii of Coaxial Cable, by Type

Cable Type	Minimum Bend Radius (inches)
RG-58/U	2.0
RG-8X	2.4
RG-8/U	4.5
RG-213	5.0
LMR-240	0.75
LMR-400	1.0

Source: NMEA

Coax Connectors

In any circuit, the weakest link is generally at a termination point, and it's no different for coaxial cable. The specialized end connectors used to install coax reflect the fact that it is really a multiconductor cable and is designed to deliver signals of varying frequencies. Our friends at the NMEA provide some guidance here as well. The maximum operating frequency and impedance for the various connector types available are shown in Table 16-6.

A few coaxial adapters. (Reprinted with permission from Boatowner's Illustrated Electrical Handbook, second edition, by Charlie Wing)

TABLE 16-6	Coaxial Connector Selection

Connector Designator	Maximum Operating Frequency	Impedance (ohms)
UHF (PL-259)	300 MHz	50
BNC	4.0 GHz	50
TNC	2.5 GHz	50
N	11.0 GHz	50
F	2.0 GHz	75
Mini UHF	2.5 GHz	50
SMA	12.0 GHz	50
SMB	4.0 GHz	50
FME	200 MHz	50

Source: NMEA

As you can see in the table, the connector impedance must match the device's operating frequency spectrum as well as the cable. Table 9-1 in Chapter 9 outlines the various equipment operating frequencies.

Installing connectors requires special care—if done incorrectly, you can create a short circuit between the center conductor and the shield. A quick Internet search will identify a plethora of specialized stripping and crimping tools available for the various coaxial cable types. They range in cost from $10 to $40, and are a good idea of you work with coax on a regular basis. Many veterans still prefer to strip the outer jacket and center conductor insulation with a pocketknife. That certainly works if you are careful, but I find the dedicated strippers quicker and more reliable (see sidebar page 173). Be aware when using a such a tool that the cutter blade orientation must be matched to the correct cable. Otherwise the tool can damage insulation or the wire conductors.

The photo sequence that follows shows the procedure for stripping cable using a specialized tool. It's important to use a tool specifically designed for the coax you are working with, because the tool's cutter spacing is designed to carefully remove insulation without damaging the conductors in the cable. (The cable stripper shown is made by Ancor and strips coax labeled RG-58, RG-59, RG-8X, and RG-62.)

You have two choices when it comes to connectors. One is the traditional soldered type, which requires you to solder the center conductor and shield to the center pin and outer case of the connector. The second type is the solderless connector (shown in the photos). Most technicians I know are moving away from the soldered type, finding them very labor intensive to install while offering no additional performance over solderless connectors. In fact, some of the newest connectors eliminate the need for the stripping procedure altogether, further simplifying this process.

ON RG-58/U, RG-59/U, AND RG-8X CABLES

STEP 1: Slip on shell and adapter; strip outer jacket back 5/8"

STEP 2: Bend back braided shield

STEP 3: Slip adapter under braided shield

STEP 4: Strip center conductor 1/2" and tin

STEP 5: Screw on body and solder tip and braid through holes in body

STEP 6: Screw shell onto body

ON RG-8/U AND RG-213 CABLES

STEP 1: Slip on shell and strip to center of conductor and back 3/4"

STEP 2: Strip outer jacket additional 5/16"

STEP 3: Slip on body, making sure shield does not contact center conductor, and solder tip and shield through holes

STEP 4: Screw shell onto body

Assembly of PL-259 (UHF) connectors. (Reprinted with permission from Boatowner's Illustrated Electrical Handbook, second edition, by Charlie Wing)

ON RG-58/U, RG-59/U, AND RG-8X CABLES
FEMALE CONNECTOR

STEP 1: Cut cable end even and strip outer jacket back 5/16"

STEP 2: Slide clamp nut and pressure sleeve over cable; straighten braid ends

STEP 3: Fold braid back; insert ferrule inside braid; cut dielectric back 13/64"; tin conductor

STEP 4: Trim excess braid; slide insulator over conductor into ferrule; slide female contact over conductor and solder

STEP 5: Slide body over ferrule and press all parts into body; screw in the nut tightly

MALE CONNECTOR

STEP 1: Cut cable end even and strip outer jacket back 5/16"

STEP 2: Slide clamp nut and pressure sleeve over cable; straighten braid ends

STEP 3: Fold braid back; insert ferrule inside braid; cut dielecric back 13/64"; tin conductor

STEP 4: Trim excess braid; slide insulator over conductor into ferrule; slide male contact over conductor and solder

STEP 5: Slide body over ferrule and press all parts into body; screw in the nut tightly

Assembly of BNC connectors. (Reprinted with permission from Boatowner's Illustrated Electrical Handbook, second edition, by Charlie Wing)

Insert the coax into the tool to the specified length.

Close the tool and rotate it 1/2 turn clockwise, then counter-clockwise, cutting through the outer insulation jacket.

Remove the cut outer jacket to expose the shield conductor.

Pull back the shield to expose the center conductor insulator.

Use the smaller-diameter cutter on the other end of the tool to strip the center conductor insulation.

The stripping process is complete and the cable is ready for the connector to be installed.

A gold-plated solderless connector made by Shakespeare (www.shakespeare-marine.com). Another popular model is available from Centerpin Technologies (www.centerpin.com). Gold plating greatly reduces attenuation due to its superior conductivity.

There are still many in the field who prefer the soldered approach but I've seen more of that type messed up than I have the appropriate solderless type. One argument for soldering is that they will be less prone to corrosion. I contend this is a moot point because when sealed as described below with 3M Scotch 2228 tape, water will not get to the connector, eliminating the corrosion issue.

Once you've made the cable termination and joined the male and female ends of the connectors, it's important to secure the connections. This is often done with Ty-wraps equipped with screw mounting brackets to relieve any potential strain on the connection. If you're installing the connection in a location exposed to weather, you will also need to seal it against water intrusion. There are several methods for waterproofing; here are two:

1 Heat-shrink tubing. This method entails dealing with the disparity in the diameters of the coaxial cable connector and the cable itself. One way to do this is by sliding two to three $1/2$-inch lengths of heat-shrink tubing (sized to accommodate the cable diameter) onto the coax before installing the connector male and female ends. Once the cable ends are attached, shrink the lengths down onto each other, which effectively builds up the outside diameter of the coaxial cable insulation. Then, when the connector is assembled (with a length of heat-shrink tubing sized

to accommodate the connector barrel length plus about $3/4$ inch on each end), the tubing will shrink down onto the previously installed short lengths, effectively sealing the entire assembly.

I don't like this method because the material only shrinks so much for its nominal size, and sometimes it's not enough to accommodate the difference in diameters between the cable and the connector.

2   Tape wrap using 3M Scotch 2228 Rubber Mastic Tape (my preference). This material seals tenaciously to itself as it's overlapped, and once in place, it is incredibly difficult to remove. In fact, a razor knife is always needed to cut it off the connection when it's time for service.

Calculating Loss

To ensure that the run from the antenna or transducer to your piece of equipment is within the tolerance specified in Table 16-2, you will need to conduct a simple loss calculation. This is an important step of any coaxial cable design/installation procedure.

To perform this calculation, simply add the attenuation through the connectors to the attenuation from the coaxial cable. The NMEA uses a constant of 0.5 dB to indicate the loss per connector for this calculation. You will also need to count the number of connections, and remember that inherent loss per 100 feet of cable run is based on the frequency.

Let's try an example using RG-8X cable to connect a VHF radio with its antenna. The radio's operating frequency is 162 MHz, which we'll round off to 160 MHz, and the cable run is 50 feet. Based on the specifications for RG-8X cable, the signal loss per 100 feet is 4.0 dB @ 160 MHz. The math looks like this:

$$(L \div 100)4 \text{ dB} = TCL$$

where L = length, and TCL = total cable loss

Plugging in the numbers above, we find that the total cable loss is:

$$(50 \div 100)4 \text{ dB} = 2$$

Then we add the loss from the two connectors (0.5 dB each), for a total loss of 3 dB. Since the maximum loss allowed by the NMEA is 3 dB for a VHF system (per Table 16-2), we're in!

VHF ANTENNAS

Like coaxial cable, antennas also greatly influence equipment performance. If the wrong choices are made in antenna selection or installation, serious problems will result.

For most boaters, their VHF radio is the all-important link to the U.S. Coast Guard if an emergency arises, so maximum radio range is an important safety consideration. Since the FCC limits power output from an installed VHF transceiver to 25 watts, the antenna installation is the key to maximizing effectiveness.

That said, many radios rated at 25 watts actually transmit at 20 or 21 watts. Lower output translates into less range at the antenna, so it is important to test for VHF power output to ensure peak transmission range. Higher-quality, name-brand radios tend to be very accurate. To perform the test, use a power standing wave ratio (SWR) meter, as outlined below:

1   Connect the meter to the VHF radio's antenna jack.

2   Plug a dummy load into the meter's antenna jack.

3   Key the transmit button on the microphone to take a direct reading of the radio's transmit power.

4   Read the wattage output on the meter's gauge. Don't be surprised if a 25-watt radio only reads 18 or 19 watts of actual output.

We'll return to SWR meters shortly.

Test rig for checking VHF power output. The connector on the left plugs into the antenna jack on the radio set, while the dummy load on the right takes the place of the antenna. (Note: Never bend coaxial cable as far as shown here. The cable was bent far beyond its specified minimum radius and held in place with a cable tie only for the purposes of the photo.)

Antenna Gain

Antenna gain is a measure of an antenna's effectiveness. It has to do with the radiated power of the antenna, but it isn't about the antenna *generating* power; it's about how the antenna *concentrates* the electromagnetic energy—the radio frequency—received from the radio. Gain is measured in decibels just like the loss factor we discussed earlier. The gain of the antenna, as well as its length, determines the radiated pattern emitted from the antenna. As the dB rating increases, the transmitting power is compressed into a progressively narrower beam, making the signal go farther, but also making its direction more critical for successful reception at the other end.

You can increase the radiated power from the VHF radio by installing an antenna with a higher gain rating, although it doesn't always improve performance in real-world conditions. A 9 dB rated antenna transmits farther, but its beam is narrower, so if it is mounted at the masthead of a sailboat and the sailboat is heeled, the signal would be angling up into the sky on one side of the boat and down into the water on the other. For sailboats, therefore, a 3 dB antenna is a more desirable choice. Even though the radiated beam doesn't extend as far as the 9 db beam, it's more likely to reach a receiving station. Powerboats, on the other hand, are inclined to run in a more level position, and they can use 6 dB or 9 dB antennas more effectively, although as a boat rolls in beam seas, transmissions may still fade in and out.

Typical radiation patterns of VHF-FM radio antennas. A 9 dB gain antenna, which might be 24 feet long, offers greater potential range but also a vertically compressed signal. When the boat rolls, you might be transmitting down into the sea on one side of the boat and up into space on the other side. Many powerboaters choose a 6 dB, 8-foot antenna as a good compromise. (Reprinted with permission from Boating Skills and Seamanship by the U.S. Coast Guard Auxiliary)

Standing Wave Ratio

The standing wave ratio (SWR), which is the ratio of the maximum RF current to the minimum RF current on the line, can be thought of as a measure of an antenna system's efficiency. (By *system*, I mean the antenna plus the coaxial cable between the antenna and the transceiver.) The smaller the ratio, the more efficient the system, so an antenna system with an SWR of 1:1 is experiencing less loss than one with a 3:1 SWR. Antenna design can have a profound effect on the SWR. By integrating various choke methods into the antenna, manufacturers can contain more RF power in the antenna element, as opposed to allowing it to leak back into the cable. You can think of the choke as a one-way valve designed to separate your antenna from the feed line, thus delivering all the power to the antenna for maximum power and directional control of the transmitted signal. Amateur radio operators call these chokes "baluns," for **bal**anced/**un**balanced, as

they balance the signal between the antenna and the cable.

To test the SWR ratio of a system, we again use the power standing wave ratio meter shown in the photo on page 175. The steps are as follows:

1 Install the meter in series between the radio transceiver and the coaxial cable routed to the antenna.

2 With the meter switched to the SWR function, key the radio microphone and observe the meter's reading.

3 Use the graph on the back of the meter to convert your reading based on the wattage output you measured (see page 174).

Align your SWR reading on the X-axis and with the Y-axis wattage reading to find the corrected SWR value.

An acceptable SWR is a matter of opinion. Some say that 3:1 is OK; others use a 1.5:1 ratio as their maximum. A value of 2:1 equals about a 90% efficiency level, and this is quite good. A 3:1 ratio represents about 75% efficiency, which, in my opinion, is a barely passing grade. A quality installation will give you a corrected reading between 1.2 and 1.6.

My Micronta (RadioShack) SWR/power meter (shown in the photos) is a long-discontinued model. Similar inexpensive meters, such as the ART-2 from Shakespeare, are available for about $50 through Boater's World stores (www.boatersworld.com). A worthwhile step up for some readers might be one similar to the Vectronics model listed in the sidebar, available from RadioShack for about $230. Some electronics specialists might benefit from a high-end unit, such as the one from Bird Electronic; however, at prices ranging from $1,200 to $1,500, these are not worth it even for advanced-level boaters.

The conversion graph on the back of the meter.

COAXIAL CABLE AND ANTENNAS

SWR METERS

ART-2, Shakespeare, www.shakespeare-marine.com
SWR-584B, Vectronics, www.vectronics.com
AT-400, Bird Electronic, www.bird-electronic.com

When considering an SWR meter, pay attention to its frequency range. For VHF use, it must able to operate in the 162 MHz frequency range. (Not all SWR meters do because most are used by CB and ham radio operators who only need a frequency range up to about 30 MHz.) The Vectronics unit has a range from 1.8 MHz to 170 MHz, and the frequency-operating range on my ancient Micronta unit is 144 MHz to 440 MHz.

SSB ANTENNAS

SSB antennas are key to the effectiveness of SSB radios—perhaps more so than with other radio installations—because the goal is very long range communication capability. Rather than being concerned with a 20- or 25-mile range, here we're thinking in terms of hundreds, and even thousands of miles! They are also more complicated, since an SSB antenna system has two parts: the abovedeck part, and the belowdeck (generally in the bilge) counterpoise part (see below). There are three types of SSB antennas commonly used on boats:

1 Vertically mounted whip antennas are common in most powerboat installations.

2 Wire (or whip) antennas are common on sailboats. The standing rigging (most often the backstay) makes up part of the antenna.

3 Halyard antennas are a new style catching on among some cruising sailors. Here the antenna is a rope halyard with a wire core of

a specific length. They are reportedly quite effective and can be raised and lowered easily when not in use.

The Counterpoise

The *counterpoise*, also called a ground plane, is a conductor or system of conductors used as a substitute for earth or ground in an antenna system. It is unique to wire or whip antennas and is a necessary component for the antenna to be effective. It allows the antenna to "see" an image of itself; i.e., to effectively reflect and transmit the radio signal. If the radiating portion of the SSB antenna system doesn't see its reflection, it simply won't work. Consider the counterpoise the second part of the antenna system. An inadequate counterpoise will most often manifest as poor range reception for the radio, especially at lower frequencies.

There are several ways to create a counterpoise:

- Use the hull of a metal boat.
- Add metal to the hull of wooden or fiberglass boats.
- Install a copper screen on the inside of a nonmetallic hull with a resin overlay to keep it in place.
- Tie into a ground plate that is in contact with the water.
- Use flat copper foil "tape" and attach it to metal through-hull fittings. (According to the NMEA, the tape should be at least 2 inches wide, but some manufacturers,

including ICOM, specify a minimum width of 3 inches.)

Because copper foil is the most common method for creating a counterpoise, we'll discuss that method in more detail.

Creating a Counterpoise with Copper Foil Tape

First let's go over a bit of background. We use foil instead of insulated copper conductors because radio frequencies do not travel *through* conductors—they travel *on the surface* of conductors. This phenomenon is sometimes referred to as the *skin effect*, and the greater the surface area, the better. One hundred square feet is often recommended as the minimum surface area for an effective ground plane. That number may seem high, but you can achieve it rather easily by tying in various bits of metal on the boat with the copper foil tape. Good choices include metal through-hull fittings,

metal water tanks, and encapsulated lead keels. The best pieces of metal are those that are located below the waterline because they will achieve an effective capacitive counterpoise to the seawater; that is, without necessarily having a direct connection to the seawater.

Warning: Never use a metal fuel tank as part of the counterpoise—you will create an explosive hazard!

To connect copper tape to a metal hull, you need an isolating connector. Use series 316 stainless steel contact points (these can be made up of 316 stainless nuts, bolts, and washers, all readily available at any chandlery), and seal the connection with a water-excluding product like Boeshield T-9 (www.boeshield.com) or Corrosion Block (www.nocorrosion.com).

If your boat is wired to ABYC electrical specifications, then the through-hull fittings will already be bonded via a green wire conductor. If so, you can route the copper tape to the

copper foil

backing plate for below-waterline ground plate

Copper tape is used to connect the SSB radio antenna to the ground plane or counterpoise. On this boat the tape is connected to a ground plate below the waterline. The metal plate shown is actually the backing plate for the Dynaplate ground plane, which is mounted outside the hull, below the waterline. Dynaplates are "sponges" made of sintered bronze. When viewed under high magnification, this material appears as a series of connected small spheres. It is extremely porous and thus exposes far more metal surface area to the water than its physical dimensions suggest.

Installation of SSB radio, antenna, and counterpoise.

SSB antenna and ground connections. (Reprinted with permission from Boatowner's Illustrated Electrical Handbook, second edition, by Charlie Wing)

Locating antennas. (Reprinted with permission from Boatowner's Illustrated Electrical Handbook, second edition, by Charlie Wing)

terminal

backstay
insulator

insulated
lead

insulator

back-
stay

The lower (left) and upper (right) connections for a backstay antenna. The insulator blocks are connected to the rod-type rigging with swageless fittings. A rigger needs to splice in these insulators at both ends. Depending on the boat's rigging type, there are various methods of splicing these insulators. Two older methods included simple ceramic insulators, with the wire rope of the stay looped through a thimble and nicopressed in place; and swaged end fittings used with cotter pins.

bonded fittings, securing them with a hose clamp or other suitable mechanical fastener, such as a stainless steel bolt and washer (if the fitting has a drilled and tapped hole).

Caution: When tying in the copper tape, avoid creating a likely scenario for galvanic corrosion (see Chapter 11). For example, never attach copper tape to an unbonded through-hull fitting. If any of the other through-hulls are bonded (as they should be), the fittings will have different ground potentials. Also, never connect copper tape directly to an aluminum or steel hull. The dissimilarity between the two different metals will create a focal point for galvanic corrosion. Remember, all that is needed to induce current flow is a difference in potential from one point to another and an electrolyte.

As you look for places to tie in, keep in mind that the counterpoise will be most effective if it is located directly below the antenna or the

loran
antennas

cell
phone
antenna

VHF
antenna

SSB
whip
antenna

A typical powerboat whip antenna is shown far right. These need to be at least 23 feet long. The wire lead from the antenna tuner to the attachment point on the antenna counts as part of the antenna's total length.

automatic antenna tuner (if one is installed), as shown in the top illustration on page 179.

Antenna Installation

As mentioned above, the two most common SSB antenna configurations are the extended whip antenna and using the standing rigging on a sailboat (see illustrations page 180). For a whip installation, the antenna needs to be at least 23 feet long; for a standing rigging installation, the antenna is typically 30 to 70 feet long, depending on the size of the boat (see photos page 181).

This variation in length may seem to contradict the precision of the SSB system and antennas in general, but by adding an automatic tuner at the base of the antenna (common on most modern boats), the effective length is automatically adjusted for the frequencies used in SSB operation. On those rare installations without automatic (or even manual) tuners, the length of the antenna must be carefully determined based on the frequencies the operator wishes to use. In reality, this is an impractical solution for any boat installation.

Safety Issues

Safety is an important consideration as you plan your SSB antenna installation. Passengers and crew must be protected against live wires and high voltage. **Keep in mind that when the SSB is in Transmit mode, the voltage delivered to the base of the antenna system can be on the order of 5,000 volts!**

Particularly with a backstay antenna, you must design the installation so that no one can reach or touch the wire accidentally during a transmission (whip antennas are typically well out of the way). At the base of the backstay rig, the height needs to be at several feet above arm's length when standing (as shown in the photo on page 185).

It really isn't a problem on the split-backstay boat shown in the photo because the connection point for the antenna is at the peak of the triangle,

well above the deck. On single-backstay rigs, you can achieve adequate protection by placing Teflon tubing over the stay, again to a height beyond arm's reach while standing.

The cable feed to the attachment point is mounted on insulated standoffs up the rig to the point where it turns into the antenna, as shown in the photo. The cable of choice for this purpose is GTO-15, which is actually ignition wire, and its insulating properties are in the 15,000 V range. Never use coaxial cable for this purpose because its insulation properties are inadequate. At the upper end of this rigged antenna, position the insulator to maintain at least 3 feet of clearance between the live wire section and the mast.

The cost and involved procedure for installing an SSB antenna in standing rigging is often one reason boatowners choose extended whip antennas for their sailboats. Whip antennas are much simpler to install. All you have to do is assemble the whip and mount it clear of other antennas by at least 4 to 10 feet (depending on the uses of the other antennas; see Table 16-9).

In summary, the considerations for SSB antenna installation are:

- Antenna length (23 feet minimum for a whip antenna; 30 feet minimum for a standing rigging antenna).
- The antenna counterpoise is critical to the range of the transceiver (more is better).
- Flat copper foil tape is the conductor of choice to connect the counterpoise.
- Position the antenna tuner and counterpoise as close to the base of the antenna as possible for maximum effectiveness.
- Use only GTO-type cable for the connection between the antenna and the tuner.
- On sailboat installations, make sure that crew and passengers can't touch the active part of the antenna during transmission.

- All electrical connections need to be first class, and insulated from weather with self-sealing tape or heat-shrink tubing.

SATELLITE TELEVISION AND RADIO ANTENNAS

We looked at some of the issues concerning GPS and DGPS antennas in Chapter 14, but we haven't yet addressed the seemingly inert television disc antennas used on many boats. These, like most of the antennas we've discussed, use coaxial cable to connect to the back of the television, so all the RFI and coiling recommendations apply here as well. These antennas typically come with 30 feet of coaxial cable attached that can be extended if necessary, as long as the loss factors are kept to a minimum.

Many disc-type television antennas also have amplification circuitry built into them to enhance reception. This circuitry requires separate DC (positive and negative) wiring and, as with all DC wiring, voltage drop should be kept to a minimum. But since this is not a critical device, a 10% voltage drop would be acceptable under ABYC standards.

You might assume that these are receive-only devices that may pick up interference from radios transmitting at frequencies close to the VHF, UHF, or FM bands. However, a few years ago the U.S. Coast Guard determined that some TV antennas can interfere with GPS antenna reception, causing inaccuracies in GPS positioning or even blinking out. Reports to the Coast Guard indicate that GPS on vessels as far as 2,000 feet from an active TV antenna have been affected. On the other hand, I've heard of cases in which interference problems were solved by increasing separation by a mere 12 to 18 inches.

The FCC identified three specific antenna models in a safety bulletin issued December 30, 2002 (TDP Electronics [i.e., Tandy Corporation] models 5MS740, 5MS750, and 5MS921), but stated that the problem may not be exclusive to these three. Similar problems were found in RadioShack model 15-1624 and Shakespeare SeaWatch model 2040 with code dates of 02A00 and 03A00.

The best way to check for a problem between the TV and GPS on your boat is to first turn on both the TV and the GPS, then shut down the TV while watching the GPS to see if any changes occur with the GPS reading or display.

Little information is available from vendors regarding mounting location, although one vendor suggests that the TV antenna be mounted as high as possible, and above all other onboard antennas. This makes sense, but it may not be feasible in many cases—for example, on a sport-fishing boat with a 30-foot SSB whip antenna. You may have to resort to the trial-and-error approach, experimenting with different locations for the TV and GPS antennas before making the installation permanent.

As of this writing, satellite radio represents one of the hottest accessories on the market. Most satellite radio antenna vendors provide a 12-foot length of coaxial cable, although this length is not crucial. And these antennas do not appear to be too sensitive as to where you locate them. As long as they have about a 90-degree clear view of the sky as measured from the center of the antenna dome (i.e., a 45-degree arc from the vertical centerline), all will be well.

CELL PHONE ANTENNAS

Most people jump on their boat and cast off without giving their cell phone's ability to make or receive calls a second thought. This generally works out because, in most cases, the body of water these people are on is actually never too far from a cell tower. But suppose you are 20 miles offshore doing a coastal delivery or anchored out at an offshore fishing ground, and you want to use your cell phone. Odds are it won't work when you're standing on a deck 1 or

2 feet above sea level with a 3-inch antenna extended out of your flip phone.

Like VHF radio and radar, cell phones are a line-of-sight system. So if we add a remote cell phone antenna, we can "stretch" the horizon out as far as possible. Like VHF antennas, these are available with different gain ratings, typically 3 dB and 9 db.

At least two companies—Shakespeare (www.shakespeare-marine.com) and Digital Antenna (www.digitalantenna.com)—make combined VHF and cellular antennas. A device splitter on the equipment end of the coaxial cable allows both systems to be attached, but of course they cannot be operated simultaneously.

The compromises already discussed in the VHF section of this chapter apply here. A 3 dB antenna generally works better at the masthead on a sailboat due to its wider beam, which helps offset the effects of the boat's heeled position. A 9 dB antenna has a longer, narrower beam. One compromise that works on both powerboats and sailboats is the use of an 8- or 9-foot whip, mounted on an extension pole set on a stern rail or on a dedicated antenna-mounting system. The antenna lead is usually terminated with a mini BNC-type plug that inserts directly into the back of the cell phone. (The socket is often hidden under a cap; you may need to pry it out.)

RADAR ANTENNAS

The most important consideration for a radar antenna or scanner—besides its output rating and whether or not it is an open or closed array (dome)—is its position on the boat. Focusing the antenna's powerful emitted beam is important to the accuracy and detail of the returned signal. The power output from the antenna or scanner is a guide to the expected range of the unit—more power means more range. Most radar antennas used on recreational boats have outputs in the 2 kW to 10 kW range, and

these are matched by the manufacturers to their respective display units. (Not all antennas are compatible with all display units, even those from the same manufacturer.)

These are powerful devices and safety must be an important consideration:

- Scanners have high-voltage components and connections within their housings. These units should only be serviced or adjusted by factory-trained technicians.

- Radar antennas transmit high-level electromagnetic energy that is dangerous to humans. They should always be mounted above head height or well below it, depending upon the boat configuration. On a sailboat, this is not a problem as the radar antenna is mounted either on a raised mounting pole far aft or well up the mast. It is sometimes a problem on a powerboat as the antenna sometimes gets mounted forward of a flybridge deck directly in front of the operator at the upper helm station. This configuration should be avoided!

- Never look directly at a scanner when it is operating, and keep people out of the emitted-beam path, which covers a full 360 degrees.

- There are strict international regulations that cover radar and its emissions. Some units that are available in the United States do not comply with European Union regulations, and may not be used there. The manufacturers usually make this information readily available. In the United States, the FCC regulates radar use; in Canada, it is regulated under the Canadian Shipping Act. As of this writing, efforts are underway to integrate Canadian and U.S. requirements.

Like VHF radio, radar is a line-of-sight device, so a higher scanner mount will give the unit greater potential range. But as with VHF, too high can be a problem for radar, as the boat's rolling and pitching is amplified at a masthead, for example. A compromise is therefore in order. The NMEA recommends a minimum height of 8 feet abovedeck, and a maximum height above the waterline of 30 feet. Keep in mind that large objects in the same horizontal plane as the radar beam will interfere with its signal and may cause blind spots or false targets to appear on the radar display.

On many sailboats, the radar scanner is mounted high on the leading edge of the mast, which may cause a blind spot on the radar screen directly aft of the boat. One way to minimize this effect is to mount a wooden block between the back of the radome and the mast, as wood is less reflective to the radar beam and will be less likely to create an obscure target on the radar screen.

On powerboats, the concern is the fore-and-aft trim angle of the boat. If you have a planing-hull powerboat, you can fabricate a wedge or bracket to adjust the antenna's plane angle to accommodate the rise of the bow when the boat is on plane (see illustrations next page). For owners of displacement-hull powerboats, this is not much of a problem since they don't experience as much bow rise underway as planing hulls.

Radar scanners are available in different power configurations. Typical small-boat units are available in the 2 kW to 5 kW power range. Larger craft often use units with power ratings up to 10 kW. Higher-output units work better in fog and rain, which is generally when a radar is most needed, because fog and mist absorb some of the radiated energy from the antenna, effectively reducing the potential range of the unit.

But radar is a line-of-sight system in both the Transmit and Receive modes. So more important than transmitting power to the system's range capability are the relative heights of the radar antenna on your boat and the target in the distance; tall is good, short is harder to see. It's worthwhile to question whether you

An effective radar mount on a sailboat. It is up out of the way of crew, elevated enough to provide an acceptable line-of-sight range, and far enough behind the mast so that the mast will interfere only minimally with its line of sight Some mounts of this type articulate automatically from side to side, leveling the antenna to compensate for the boat's heeling angle. (This photo also shows the split-backstay SSB antenna installation discussed on page 182.)

really need a radar with a 50-mile range, which the vast majority of boaters never use. As a boater, I'm more concerned with seeing targets as they get closer to me so that I can avoid hitting them! Rather than impressing people with a 10 kW, 50-mile radar, I'd rather

When mounting a radar antenna on a powerboat, ensure that the 25-degree vertical beam is leveled horizontally and not obstructed by structures on the boat itself. Adjustable brackets and shim kits are available from radar suppliers as aids in leveling the unit (inset).

A proper radar installation on a 35-foot lobster boat. Notice the slight downward tilt to the radome relative to the cabin top. When the boat is running and its bow rises, this radome will be level, or parallel to the water's surface, which is the goal.

have a unit with adjustable intrusion alarm features and MARPA (mini automatic radar plotting aid) capabilities. And for the average boater, a 4 kW unit will do the job nicely.

Although radar is a line-of-sight system, its range is actually a bit farther than that. Because atmospheric refraction bends the beam slightly, a radar can actually "see" about 15% farther than a human observer located at the same height. The location of the antenna on the boat (mounting it higher extends the horizon), the height of the target, and the visible horizon all make a difference. The height of targets located beyond the visual horizon determines the actual radar range. This distance (see Table 16-7) can be calculated as follows:

$$D = 1.144\left(\sqrt{H_1} + \sqrt{H_2}\right)$$

TABLE 16-7	Radar Line-of-Sight Range in Nautical Miles								

| Radar Height | Target Height (H_2, ft.) | | | | | | | | |
(H_1, ft.)	5	10	20	40	60	80	100	200	400
5	5	6	8	10	12	13	14	19	26
10	6	7	9	11	13	14	15	20	27
15	7	8	10	12	14	15	16	21	28
20	8	9	11	13	14	16	17	22	29
25	9	10	11	13	15	16	18	22	29
30	9	10	12	14	16	17	18	23	30
40	10	11	13	15	17	18	19	24	31
50	11	12	14	16	17	19	20	25	32

Reprinted with permission from *Boatowner's Illustrated Electrical Handbook*, second edition, by Charlie Wing

where D = distance in nautical miles,
H_1 = height in feet of the antenna, and
H_2 = height of the distant object

Increase this distance by 15% for the absolute maximum radar range.

While the height of the radome has an obvious effect on maximum range, it affects minimum range as well. The radiated beam from the radar transmitter extends in a vertical arc 12.5 degrees above and below the horizontal centerline. The area below the lower half of the beam is a radar blind spot, or blind zone, and the higher the antenna is mounted, the farther out this blind spot will extend (see Table 16-8). The near distance can be calculated from:

$$N_1 = H_1 \div \text{tangent } 20° = 2.75 \times (H_1 - H_0)$$

For example, a 40-foot-high antenna cannot see surface objects closer than 110 feet. Reducing the antenna height to 20 feet reduces the blind zone radius to 55 feet. On most boats, practical mounting locations are fairly limited, however, so the near blind spot may be relatively fixed. That said, I feel that radomes mounted on purpose-built articulating pole mounts, like the one shown in the

Radar range and blind zone versus heights of the antenna and the target. (Reprinted with permission from Boatowner's Illustrated Electrical Handbook, second edition, by Charlie Wing)

TABLE 16-8	Radar Near Range (Blind Zone), N_1, in Feet								

Radar Height (H_1, ft.)	Target Height (H_0, ft.)								
	0	1	2	3	5	10	15	20	30
5	14	11	8	5	—	—	—	—	—
10	27	25	22	19	14	—	—	—	—
15	41	38	36	33	27	14	—	—	—
20	55	52	49	47	41	27	14	—	—
25	69	66	63	60	55	41	27	14	—
30	82	80	77	74	69	55	41	27	—
40	110	107	104	102	96	82	69	55	27
50	137	135	132	129	124	110	96	82	55

Reprinted with permission from *Boatowner's Illustrated Electrical Handbook*, second edition, by Charlie Wing

TABLE 16-9	Recommended Horizontal Antenna Spacing (in feet, with vertical spacing where required)

	VHF	GPS	SSB	Radar (4 kW)	Satellite Communications	Cell Phone	Satellite TV
VHF	4	3	3	2	6	2	3
GPS	3	0.5	4	vertical separation; GPS out of radar beam	6 (and vertical separation; GPS below satcom)	5	2
SSB	3	4	10	3	6	2	4
Radar (4 kW)	2	vertical separation; GPS out of radar beam	2	1.5 vertical separation	6	cell out of radar beam	4 (more for higher-power radar)
Satellite Communications	6	6 (and GPS below satcom)	6	6	6	6	6
Cell Phone	2	5	2	cell out of radar beam	6	3	4
Satellite TV	3	3	4	4 (more for higher-power radar)	6	4	4

Source: NMEA

photo on page 185, offer the best compromise between serviceability, close range, and long-distance capabilities.

With so many different kinds of antennas on board, it is almost inevitable that issues of interference will occur. However, by closely following manufacturers' recommendations concerning antenna location relative to other transmitting and receiving antennas, you can mitigate most of these issues.

Table 16-9 summarizes the recommendations of various manufacturers, but you should refer to each product's documentation for the final say.

Onboard Networks

The marine industry is going through perhaps the most significant change in thirty years regarding onboard systems. Boaters are beginning to embrace what has been commonplace in our working lives for some time now—the networking of electrical and electronic components—with the result that sending data from one part of the boat to another is becoming mainstream. Many systems or methods, each with their own strengths and weaknesses, are vying for market share and technical acceptance. Vast changes will occur in boat wiring in the near future; changes that may, to a large extent, do away with wiring altogether. Wireless systems are even being installed on a few very high end boats today, although widespread market penetration is, I believe, still several years away.

How will boatowners and marine electricians deal with all of this? What skills will be needed to remain effective and relevant in servicing and installing these systems? These are the questions I'll address in this chapter. To facilitate our networking discussion, see the sidebar opposite on networking terminology, which includes the terms introduced in Chapter 14.

NETWORK TYPES

Networking is not new to boats; the NMEA 0180 network standard—the first communications standard created for the marine industry—has been around since 1980, and was followed shortly by the 0181 standard, and then 0183, which became widely accepted and was the mainstay for nearly twenty years. NMEA standards are accepted as the common practice for how to share specific data information between competing entities or companies.

But over the years we've become so used to having nearly unlimited amounts of information at our fingertips at all times and in all places that advancements in this area were all but inevitable. Today, networks and proprietary systems are all the buzz. Some use twisted-pair harnesses to eliminate the effects of EMI; others use shielded wire sets. But the field problems related to networking go far beyond a difference in cable types. Issues such as data transfer rates, the types of data, plug connector compatibility, the number of devices that can be connected to the network, and the compatibility of equipment from one vendor to another are concerns as well.

Let's look at several networks to get a feel for what's available and how they work.

Furuno NavNet

Furuno's NavNet system is an Ethernet-based network, specifically a 10Base-T system. Ethernet protocol is available in several configurations that allow for varying amounts of data to be transmitted. 10Base-T is based on an IEEE standard protocol for identifying system features. In this case, "10" simply refers to the data transmission speed in megabits (Mb)

COMPUTER AND NETWORKING TERMINOLOGY

In this discussion of networks, we will be entering what may be for some a new realm of specialized knowledge. To assist you, here are definitions for some of the terms used in this book:

bandwidth: A range within a band of frequencies or wavelengths. Also, the amount of data that can be transmitted in a fixed amount of time; the data rate supported by a network connection or interface. Commonly expressed in bits per second (bps).

baseband: The original band of frequencies of a signal before it is modulated for transmission at a higher frequency. Typically used to refer to the digital side of a circuit when the other side is broadband or frequency based, meaning the signal is modulated. For example, in a cell phone, "RF to baseband" means pulling out the analog or digital data from the modulated RF signal received from the cell phone tower and converting it to pulses for digital processing.

black box: A sealed, unserviceable electronic control box that serves various functions within an electronic or networked system.

broadband: In general, broadband refers to telecommunication in which a wide band of frequencies is available to transmit information.

CAN (controller area network): A serial bus network of microcontrollers that connects devices, sensors, and actuators in a system or subsystem for real-time control applications. There is no addressing scheme used, as in the sense of conventional addressing in networks (such as Ethernet). Rather, messages are broadcast to all the nodes in the network using an identifier unique to the network. Based on the identifier, the individual nodes decide whether or not to process the message and also determine the priority of the message in terms of competition for bus access. This method allows for uninterrupted transmission when a collision is detected, unlike Ethernets that will stop transmission upon collision detection. Controller area networks can be used as an embedded communication system for microcontrollers as well as an open communication system for intelligent devices.

Ethernet: Originally developed by Xerox in 1976, it now describes a diverse family of frame-based computer networking technologies for local area networks (LANs). It is the basis for IEEE standard 802.3. (If you want to learn more about Ethernet protocol, the Internet is absolutely loaded with sites that will walk you through the various protocols and the cryptic codes that identify them.)

multiplexer: A communications device (black box) that combines several signals for transmission over a single medium.

network: A group of two or more computer systems linked together to share information and hardware.

network architecture: The structure of a communications network. An *open architecture* allows adding, upgrading, and swapping of components. It can be connected easily to devices and programs made by other manufacturers. It uses off-the-shelf components and conforms to approved standards. A *closed architecture* has a proprietary design, making it difficult to connect the system to other systems. The hardware manufacturer chooses the components, and they are generally not upgradable.

protocol: In the context of data communication, a common set of rules, signals, and data structures (for either hardware or software) that governs how computers and other network devices exchange information over a network.

sentence structure: In NMEA parlance, describes the sequence and type of data information codes that are distributed throughout the network. (For specific sentence codes, see the sidebar on page 151.)

per second. "Base" describes the transmission type (in this case, baseband as opposed to broadband). The "T" describes the wire type (in this case, unshielded twisted-pair cables). (In contrast, the protocol also describes a 10Base-F system that utilizes fiber optics as the transmission medium.)

The NavNet system uses proprietary language protocols and is only compatible with Furuno products, so mixing and matching among vendors to create an onboard system will be challenging. Some Furuno products are also at least partially NMEA 0183 compatible, which will allow for some interfacing of non-Furuno products, but this interface capability is limited. The NavNet system is essentially a closed, proprietary system.

Xantrex Xanbus

The Xanbus network system, developed by Xantrex Technology, is a proprietary system. It is based on CAN bus, and is compatible with NMEA 2000. It uses standard computer-type Cat 5 cabling to connect its various components and readily available RJ-45 terminal connectors, familiar to anyone who has worked around computers or office networks. While this system uses some NMEA 2000 language for data communications, it does not use NMEA 2000 proprietary cabling to connect its various components. So again, we see that nothing in this system is really standardized to allow simple integration of components from different vendors.

Mercury SmartCraft

SmartCraft is based on CAN technology to integrate its electronically controlled engine products with other onboard systems such as engine monitoring instruments, generator remote-start panels, and the like. Since CAN has been used in automotive applications since the late 1980s, this was a logical move for Mercury because its applications are centered on automotive-based engines that have been marinized, and the automotive market is certainly much larger than the marine market, with lots of experience in the utilization of CAN in car and truck applications. So in this case, the logic was to go with a proven technology. Again, mixing and matching SmartCraft equipment with other networking protocols is not easily accomplished.

Raymarine SeaTalk

Raymarine uses its own proprietary SeaTalk network system, which can transfer data at 100 Mb per second. There have been many iterations of the system over the years as equipment needs have increased. Raymarine also embraced NMEA protocols, producing both NMEA 0183- and NMEA 2000-compatible equipment. In the past, though, integration with other NMEA equipment from other vendors has required the use of black boxes to translate and distribute information from Raymarine's protocol to the open architecture of the various NMEA standards.

NMEA 2000

In an effort to establish a universal approach to networking, the NMEA developed NMEA 2000 so that a true "plug and play" protocol could be established. The NMEA 2000 standard is designed to allow equipment from different manufacturers to interface with each other—as long as they are compliant with NMEA 2000.

The problem is that in spite of this noble effort, NMEA 2000 is still flawed. There are several reasons for this. One is that the NMEA included a requirement that equipment be NMEA 2000 certified—at a hefty fee, especially for equipment vendors who have already invested heavily in an alternative protocol. Some feel that the expense involved is too much of a burden, which may explain why marine electronics equipment manufacturers are not jumping on the NMEA 2000 bandwagon. In fact, as of this writing, the NMEA 0183 standard is still widely used.

Further, in spite of the significant improvement in data transfer rates (up to 250,000 bps per second versus 4,800 bps for the earlier 0183 system, or at best 38,400 bps for the high-speed version of NMEA 0183), the system still can't handle the system requirements of some of the latest equipment.

It should also be noted that this 250,000 bps data transfer rate is based on a 200-meter-long network. Theoretically a 25-meter-long CAN bus—a more typical length for many recreational boats—could transfer data at as much as 1,000,000 bps per second (1 Mb), but the NMEA 2000 standard has fixed the value at 250,000 bps.

One example of how this thinking may be flawed is with video data transfer. On a modern boat, it's quite conceivable that an owner may want to have satellite TV broadcasts available at the helm station, visible on the same screen on which the radar image and chartplotter displays appear. Another example is video monitoring of an engine room area, a common practice on larger yachts. NMEA 2000, which like the SmartCraft system is based on CAN protocol, can't handle the data load these systems require because it doesn't have enough bandwidth.

Typically, vendors revert back to an Ethernet-based addition to the NMEA 2000 system to accommodate needs of this nature. Depending upon the version of Ethernet used, data transfer rates can be as high as 1 gigabit (Gb) per second. So, in the end, we still see installations that use two or more network protocols to accommodate the needs or desires of today's boatowner.

TROUBLESHOOTING

The bottom line is that in spite of efforts to standardize data transfer, no one is there yet, and this puts the burden on systems and network installers to be really up to speed with the equipment from the vendors in question. The only way to achieve this is through factory training, which is not something a boatowner, or even a typical marine electrician, has access to, nor would such a high level of product-specific knowledge make sense (especially as it becomes obsolete so quickly).

In the end, about all a boatowner or marine electrician can expect to be able to do with any of these networked systems is to confirm the integrity of the electrical connections and wiring that tie the system components together. It doesn't matter if the network is wired with twisted pairs, CAT 5 cable, or a shielded harness; they are all basically a group of electrical conductors inside a sheath. In other words, although components are linked with specialized data cables made of very small gauge wire (typically 18 AWG and smaller) and special connectors with tiny contact points, it is still "just wire." And while this very small wire is primarily carrying data, it is often also used for low-power distribution.

So many of the troubleshooting methods I've described up to now apply here as well. We still need to confirm the following:

- all the points in the circuit are properly connected
- there is no signal degradation due to EMI or RFI
- excessive voltage drop does not exist where electric power is being distributed through the network cabling
- appropriate overcurrent protection is provided

Beyond these considerations, the systems of today are really "black boxes" in the sense that the high-tech side of things is literally "in the box" and not field serviceable. As a troubleshooter, your role may consist of nothing more than being able to trace and confirm that the connections are in the proper order and there are no breaks in the length of the wire running between points A and B. As with shore-power AC systems and traditional DC-distribution systems with their relatively large-gauge cabling, the vast majority of problems in data networks will almost always be "at the ends," or termination points. (There is also the possibility of physical damage to a cable run, but this will often be obvious from a simple visual inspection.) Your tools of choice will be:

- TDR
- tone-generating circuit tracer
- ohmmeter
- voltmeter

Here are some basic guidelines:

- Use a TDR or tone-generating circuit tracer (described in Chapter 3) to trace the cables and confirm the integrity of the wire run. (You can also use an ohmmeter for this task.)

  If you use a TDR, remember its minimum operational length requirement is 9 feet (2.7 m). Many of the cable lengths you might be testing are less than 9 feet, so you will have to add a 9-foot length of cabling to one end of the cable being tested to get relevant readings on the TDR. An easier and more practical option is to use a tone-generating circuit tracer or ohmmeter to "pin out" the harness.

- Use a voltmeter to test that any distributed power is within acceptable limits.

- Look for solid continuity from a pin terminal at one end of the harness assembly to the corresponding pin at the other end. Do a wiggle test (see Chapter 15) at the plug assembly with the meter attached. Wiggle the plug assembly while observing your meter to see if the wire being tested alternates between continuity and open circuit. Remember almost all harness failures occur at the plug ends.

- Confirm that the various wires are connected to the correct terminals, and that the individual wires have good continuity. Operational power for the various pieces of equipment in most networks has historically been provided separately by the traditional DC positive and negative wiring connections to each piece of gear, along with grounding connections for some equipment. This concept is changing rapidly, however, and in fact, both SmartCraft and NMEA 2000 network cables are also used for power distribution.

- Confirm power delivery through the common power distribution/data delivery cable harness. This is relevant if an NMEA 2000 network is being used to supply low-current power to the connected equipment in addition to transferring data. The limit for power distribution under the NMEA 2000 standard is 1 amp per device connected, if needed. For devices requiring more than 1 amp, auxiliary power supply lines are required. The NMEA 2000 proprietary cables used in such a system are available in two configurations, described as heavy and light cable. The heavy cable is intended for cumulative loads of up to 8 amps, and the light cable is designed to handle cumulative loads of up to 4 amps.

Signs That Troubleshooting Is Necessary

Signs of problems with networked equipment include dim screen displays and "scrambled" or nonexistent data. If the equipment has a

An NMEA 2000 network can distribute up to 1 amp per device over network wiring. Beyond that, dedicated power leads are required.

record of satisfactory performance, do not immediately assume the networked equipment is at fault. The first step is to determine if the more vulnerable wired connections are faulty. If you have a new installation, the problem may be software incompatibilities or glitches in the software itself. (Whenever I see that a piece of gear is running or controlled by version 1.0 of any software, I'm always suspicious. The programming people don't always get things perfect on the first go-round.) If you suspect a software problem, call the equipment dealer or manufacturer to find out if there is a history of problems, and what the software solution might be.

The next step is to positively identify the function of each of the networking cables connected to the electronic equipment, and

then determine the function of each of the pin connections in the cable assembly. You'll need to refer to the equipment owner's manual or installation guide for this information. If you don't have the documentation, you might as well stop. You must know what role each and every wire and pin play in the system to determine if all the "dots" are connected properly and that the cable harness is in good working order.

The NMEA installation standard clearly identifies standard color codes for its 2000 standard cable harnesses. These colors apply to the individual conductors within the multi-conductor cable. Due to the cable design, however, you won't really be able to see the various colors; you are going to have to work from pin to pin at the cable ends to confirm proper continuity of conductors.

Remember that failures at the connection point on the equipment are not uncommon, and with a visual inspection you'll be able to see corrosion or loose wires. Slight corrosion has a much more profound effect on a network system's ability to distribute data than on more traditional electrical connections, so carefully inspect the cables and connections. Look for any signs of corrosion on the pins or socket ends of the termination points. If you see loose wires at screw and ring-eye connections, tighten them.

If you use an ohmmeter for these tests, take into account the inherent resistance of the wire, which is very small gauge, typically 16 to 24 AWG (1 to 0.2 mm^2). The NMEA 2000 standard identifies maximum allowable resistance values for power runs as 1.34 ohms per 100 meters for heavy-gauge network cable (16 to 18 AWG) and 5.4 ohms per 100 meters for light-gauge network cable (22 to 24 AWG). Table 17-1 identifies typical resistance values for commonly used copper wire gauges as well as the metric equivalents.

TABLE 17-1	Resistance for Copper Wire				
AWG	Feet/ohm	Ohms/ 1,000 feet	mm²	Meters/ohm	Ohms/ 1,000 meters
10	980.4	1.01	5.26	304.8	3.28
12	617.3	1.62	3.31	191.9	5.21
14	387.6	2.58	2.08	120.6	8.29
16	244.5	4.09	1.31	75.9	13.17
18	153.6	6.51	0.82	47.7	20.95
20	96.2	10.4	0.52	30.02	33.31
22	60.6	16.5	0.32	18.9	52.95
24	38.2	26.2	0.21	11.9	84.22
26	24.0	41.6	0.13	7.47	133.9
28	15.1	66.2	0.08	4.69	213.0

Specifications are for copper wire at 77°F (25°C).

NMEA 2000 also specifies maximum allowable voltage drop as supplied by a battery at 1.5 V, and it goes into great detail explaining step-by-step procedures for checking each leg of the network as more nodes are added to the system. Use the inTELLECT meter (see Chapter 2) to determine if cabling is within the allowable parameters. Concerns with voltage drop apply to any network system, so always identify and use specific manufacturers' requirements. If you really want to learn more about the intricate detail of these systems, I strongly recommend that you get a copy of the NMEA's *Installation Standards for Marine Electronic Equipment Used on Moderate-Sized Vessels*, second edition. Go to www.nmea.org to order a copy (it's not free—it's $175 for NMEA members and $295 for non-members—but it's worth every penny).

THE FUTURE OF ONBOARD POWER DISTRIBUTION

Another area of onboard wiring that is rapidly changing is power distribution to equipment other than electronics. This applies to DC and,

in some cases, AC power. Many companies have entered the marketplace, some of them with systems that have been used successfully in the commercial aircraft industry for several years. These same systems also use computer-based protocols to send data, and as already stated there are protocols in place other than the NMEA 2000 standard. Some are CAN-based networks just like NMEA 2000 and SmartCraft systems, and others are not.

Three-Wire Power Distribution

The basic goal behind these new power distribution systems is to dramatically reduce the actual amount of wiring needed on board the boat. The systems will still use the traditional DC positive conductors and negative returns to and from a main distribution panel. The essential difference, though, is that these systems will use only one positive conductor and one negative conductor to power multiple DC loads distributed throughout the boat. The third wire in the system will be used as a two-way data communication cable that will have multiple uses, such as actual data

A simplified view of a three-wire (positive, negative, and data) power distribution installation. It's important to note that not all of the available distributed power systems emerging as of this writing are actually three-wire systems. In fact, an NMEA 2000 system, which distributes both data and low-level power, uses five wires in its standard cabling. Some SmartCraft cabling can incorporate as many as eight conductors in its harness assemblies.

A distributed power system from Capi2 (www.capi2.com). The large-gauge red and black cables are the power distribution conductors, and the smaller center wire is the data transfer cable. The box the cables are running through is the local distribution point for a given piece of equipment. Any additional cabling would be tapped from this box to the specific device.

A centralized control panel/monitor from Moritz Aerospace (www.moritzaero.com). This electronic touch-screen panel monitors all the data and power distribution circuits connected to it.

transmission for equipment monitoring and for switching the device on and off. By having the primary power delivery system shared among various components, it's fairly easy to see how a designer could save on the number of wires in the overall system. This not only simplifies the system by reducing the number of actual wires needed, it ultimately can be translated into a huge reduction of weight in copper on board a boat. Considering that the price of copper has skyrocketed over the last few years, this whole concept can ultimately save money just in the material requirements used to make up a system.

CLOSING THOUGHTS

Industry standards have yet to be established for things like circuit protection and environmental installation concerns for distributed power

systems such as these, although the ABYC is working on them diligently. And those of us who work within the industry are still learning about these emerging systems along with the problems and issues related to them. With the introduction of each system, new questions about compliance with U.S. Coast Guard regulations and existing ABYC standards get raised.

As for which system is better or worse, my opinion is that it's too difficult to determine at this time. All the systems we have covered have strengths and weaknesses. Systems vary in their amount of bandwidth, transfer capabilities, effective operational length, and the number of devices they can incorporate into their network. And equipment vendors have yet to agree on how to standardize the various networking options.

That said, the way we check for the age-old problems has indeed advanced, and microprocessor-driven devices make the process of finding the old problems faster, easier, and more accurate than ever before. Once you've eliminated problems associated with cables and connectors, it's all "in the box," and there is little or nothing that a boatowner or marine electrician can do in a field situation to correct such problems. Production of electronic equipment has "evolved" to the point where servicing internal components is not cost effective or even possible.

As an example, my microwave oven recently stopped working. As an experienced electrician, I figured it wouldn't be a problem to fix it, and dove in headfirst. After disassembling the oven, I was sure I'd located the faulty component—the integrated circuit board. I ordered a new part from a local appliance dealer, and ten days later, it arrived—at a cost of $200. I installed it . . . and the microwave still didn't work. My next stop was a Best Buy, where I bought a new microwave that did more than my old unit, and cost $300. For only $100 more initially, I could have saved myself several days of labor and ordering and waiting for parts. Instead, I shelled out $500 plus my labor and time.

Most of the equipment on today's boats falls into the same category as my microwave. If the device is still covered under warranty, great; the company will give you a new unit. If not, it may be worth sending it to the factory for a look, but that will depend on the overall cost of the device.

On the upside, vendors are now able to pack more intelligence and functionality into equipment than was imaginable ten years ago. And, according to several manufacturers I've asked, the failure rate is very low—somewhere between 1% and 5% of total units sold. To me, the convenience and capability of the gear we buy and use today is well worth the occasional failure.

Metric Equivalents

Multiply	English to Metric By	To Get		Multiply	Metric to English By	To Get
LENGTH						
Inches	25.4	Millimeters		Millimeters	0.0394	Inches
Inches	2.54	Centimeters		Centimeters	0.3937	Inches
Inches	0.0254	Meters		Meters	39.37	Inches
Feet	30.48	Centimeters		Centimeters	0.0328	Feet
Feet	0.3048	Meters		Meters	3.281	Feet
Yards	0.9144	Meters		Meters	1.094	Yards
Miles	1.609	Kilometers		Kilometers	0.6215	Miles
AREA						
Inches^2	645.16	Millimeters^2		Millimeters^2	0.00155	Inches^2
Inches^2	6.4516	Centimeters^2		Centimeters^2	0.155	Inches^2
Feet^2	929.03	Centimeters^2		Centimeters^2	0.00108	Feet^2
Feet^2	0.0929	Meters^2		Meters^2	10.764	Feet^2
Yards^2	8361.3	Centimeters^2		Centimeters^2	0.00012	Yards^2
Yards^2	0.8361	Meters^2		Meters^2	1.196	Yards^2
Miles^2	2.59	Kilometers^2		Kilometers^2	0.3861	Miles^2
VOLUME						
Inches^3	16.387	Millimeters^3		Millimeters^3	6.1×10^{-5}	Inches^3
Inches^3	16.387	Centimeters^3		Centimeters^3	0.061	Inches^3
Feet^3	0.0283	Meters^3		Meters^3	35.33	Feet^3
Yards^3	0.7646	Meters^3		Meters^3	1.308	Yards^3
ENERGY						
Ergs	10^{-7}	Newton-meters		Newton-meters	10^7	Ergs
Joules	1	Newton-meters		Newton-meters	1	Joules
Joules	10^7	Ergs		Ergs	10^{-7}	Joules
Joules	0.2389	Calories		Calories	4.186	Joules
Joules	0.000948	Btu		Btu	1,055	Joules
Joules	0.7376	Foot-pounds		Foot-pounds	1.356	Joules
Calories	0.00397	Btu		Btu	252	Calories
Joules/see	3.41	Btu/hr		Btu/hr	0.293	Joules/see
Btu/hr	252	Calories/hr		Calories/hr	0.00397	Btu/hr
Horsepower	746	Watts		Watts	0.00134	Horsepower
F°	0.556	C°		C°	1.8	F°
°F	0.556 (°F − 32)	°C		°C	1.8 x °C + 32	°F

Reprinted with permission from *Boatowner's Illustrated Electrical Handbook*, second edition, by Charlie Wing

Resources

I'm often asked about websites that are useful to the marine electrician or boatowner for keeping up to date on new developments, as well as sourcing the equipment mentioned throughout this book. So below are some of my most frequently visited websites; some are mentioned in the text, but others are just good resources for further information.

WEBSITES

AEMC Instruments, www.aemc.com. Manufacturer of highly specialized meters, such as time domain reflectometers (TDRs). Also offers power analyzers and software for advanced electricians.

Airpax, www.airpax.net. Manufacturer of circuit breakers and ground fault circuit interrupters, among other things.

American Boat and Yacht Council, www.abyc.com. A great site to learn about the standards that apply to electrical work, changes to standards, and educational programs available to both boaters and tradespeople.

Ancor Marine, www.ancorproducts.com. A good source for learning about available wire, cable, and terminals, and for ordering specialized tools used in the trade, including the Wire Tracker tone-generating circuit tracer.

Blue Sea Systems, www.bluesea.com. A good source for electrical panels, fuses, circuit breakers, and switches. Also offers excellent installation education materials.

Carling Technologies, www.carlingswitch.com. A great site to learn about circuit breakers and their characteristics.

Centerpin Technology, www.centerpin.com. Manufacturer of terminal connectors for coaxial cable and battery cable, battery clamps, and other wiring hardware.

Charles Industries, www.charlesindustries.com. A great source for inverters, regulators, and other devices discussed in the book.

E-T-A, www.e-t-a.com. A good resource for learning about circuit protection device characteristics.

Extech Instruments, www.extech.com/instrument. Offers a comprehensive selection of test equipment, including infrared thermometers.

Fluke, www.fluke.com. The manufacturer of some of the best test instruments and software solutions available to the marine electrician, including the oscilloscopes discussed in this book.

Ideal Industries, www.idealindustries.com. Manufacturer of the SureTest circuit analyzer, circuit tracers, power quality meters, and other test equipment.

Midtronics, www.midtronics.com. Manufacturer of the Micro500XL battery and electrical system analyzer and the inTELLECT EXP-1000 diagnostic platform. This site also provides details on how conductance testing works.

Midwest Corrosion Products, www.no corrosion.com. Manufacturer of Corrosion Block, a corrosion inhibitor and lubricant.

Mytoolstore, mytoolstore.com. This site offers good prices on some of the test equipment mentioned in this book.

National Marine Electronics Association, www.nmea.org. The NMEA site provides access to its standards and information on its educational programs. You can purchase the installation standards here as well.

PMS Products, www.boeshield.com. Manufacturer of Boeshield T-9, a corrosion protectant and lubricant.

***Professional Boatbuilder*, www.proboat.com.** Website for the magazine. For the professional who wants to stay up to date, it is the source for the latest in technological developments in all areas of marine technology.

Professional Equipment, www. professional equipment.com. A very good source for many of the test meters mentioned in this book, including the WattsUp? portable watt meter from Electronic Education Devices (www.doubleed. com).

ProMariner, www.promariner.com. A good source for the electrical gear mentioned in this book as well as some test equipment specific to cathodic protection.

PSICompany, www.psicompany.com. Retailer for the Newmar StartGuard DC Power Conditioner and other power supplies and conditioners.

Siemens Water Technologies, www.usfilter. com. Manufacturer of cathodic corrosion protection systems, including silver chloride reference electrodes.

Sperry Instruments, www.awsperry.com. This site offers a wide range of telecommunications and electronics test equipment.

Tequipment, www.tequipment.net. Retailer of testing equipment from many manufacturers.

Yokogawa, www.yokogawa.com. Manufacturer of high-resolution AC leakage amp clamps.

ABYC Standards of Interest for Marine Electricians and Boatowners

The ABYC standards are mentioned quite frequently in this book. I'm often asked, "What standards should I be really familiar with to be better at my work as a marine electrician?" Well, here's the list:

- A-27, Alternating Current (AC) Generator Sets
- A-28, Galvanic Isolators
- A-31, Battery Chargers and Inverters
- E-2, Cathodic Protection
- E-10, Storage Batteries
- E-11, AC & DC Electrical Systems on Boats
- P-24, Electric/Electronic Propulsion Control Systems
- TA-27, Batteries and Battery Chargers (technical information report)
- TE-4, Lightning Protection (technical information report)

If you use these standards, it is imperative that you understand that they are "dynamic documents," which means they are constantly evolving. Therefore, be sure you are working with the latest version of the standard; otherwise, you may miss out on important changes.

To acquire the standards, you must become a member of the ABYC. As a member, you'll receive regular updates on the work of the technical committees on new and revised standards. To find out more, go to www.abyc.com.

GLOSSARY

Absorbed glass mat (AGM) battery: A lead-acid battery in which the electrolyte is held in fiberglass plate separators. The mat, which separates the battery plates, is saturated with electrolyte, effectively immobilizing the liquid.

AC ripple: A measurable amount of AC current that will leak through a typical rectifier circuit designed to convert pure AC to DC. Excessive AC ripple can be a source of electronic noise. It can also cause batteries to overheat if they are connected to a battery charging source that has excessive leakage.

American Wire Gauge (AWG): A U.S. measurement standard for the diameter of non-ferrous wire, which includes copper and aluminum. The smaller the AWG number, the thicker the wire, and vice versa. Although this seems counterintuitive, it is easily explained. In making wire, metal is pulled through a series of increasingly smaller dies to create the final wire size. The AWG number is the number of dies. The more dies, the larger the number and the smaller the diameter.

Ampere interrupting capacity (AIC): An important fuse or circuit breaker rating that describes the amount of amperage the device can be exposed to and still function as a circuit interrupter.

Anode: The negative electrode in a galvanic cell. The positive terminal in an electrolytic cell.

Antenna gain: A measure of the effectiveness of a directional antenna as compared to a standard nondirectional antenna. Also called "gain."

Apparent power: The total electrical energy actually delivered by the power supply (i.e., a utility company, an AC generator, or a DC-to-AC inverter).

Attenuation: The amount of signal loss per given length of the cable.

Available short-circuit current (ASCC): How much current can flow through a circuit in spite of resistance factors. Current available beyond a circuit breaker's interrupting capacity.

Bandwidth: A range within a band of frequencies or wavelengths. Also, the amount of data that can be transmitted in a fixed amount of time; the data rate supported by a network connection or interface. Commonly expressed in bits per second (bps).

Baseband: The original band of frequencies of a signal before it is modulated for transmission at a higher frequency. Typically used to refer to the digital side of a circuit when the other side is broadband or frequency based, meaning the signal is modulated. For example, in a cell phone, "RF to baseband" means pulling out the analog or digital data from the modulated RF signal received from the cell phone tower and converting it to pulses for digital processing.

Black box: A sealed, unserviceable electronic control box that serves various functions within an electronic or networked system.

Bonding system: Protects a boat against stray-current corrosion by electrically tying together all the metals on the boat and then connecting the bonding wires to the boat's common ground point.

Broadband: In general, broadband refers to telecommunication in which a wide band of frequencies is available to transmit information.

Capacitance: The ratio of the electric charge transferred from one to the other of a pair of conductors to the resulting potential difference between them.

Capacitor: An electronic component used for storing charge and energy. The usual capacitor is a pair of parallel plates separated by a small distance. When a steady voltage is applied across a capacitor, a charge $+Q$ is stored on one plate while $-Q$ is stored on the opposite plate. The amount of charge is determined by the capacitance (C) and the voltage difference (V) applied across the capacitor: $Q = CV$. Since the charge cannot change instantaneously, the voltage across a capacitor cannot change instantaneously either. Thus capacitors can be used to guard against sudden losses of voltage in circuits, among other uses.

Cathode: The positive electrode in a galvanic cell. The negative terminal in an electrolytic cell.

Cathodic protection: Protects a boat against galvanic corrosion through the use of zinc anodes.

Coaxial (coax) cable: A cable consisting of four components: (1) a center conductor made of either solid copper wire or stranded copper (commonly used in marine applications; (2) a dielectric material or insulator surrounding the center conductor; (3) a second conductor, which is a sheath or shield made of either braided copper or a foil; and (4) an outer jacket that protects the cable from the environment and provides some flame retardation. Coax is used primarily for transmission of high-frequency signals.

Code of Federal Regulations (CFR): Rules and regulations established by the federal government, covering a broad range of governmental branches (e.g., the U.S. Coast Guard) and topics, including the installation of boat systems (Titles 33 and 46). The regulations apply to both recreational and commercial boats.

Cold cranking amps (CCA): One of several battery ratings. The number of amps that a new, fully charged battery at 0°F (–17.8°C) can deliver for 30 seconds and still maintain a voltage of 1.2 or more volts per cell.

Compass deviation: Compass error induced by magnetic interference (such as produced by an electrical current). The amount of error will vary proportionally based on the strength of the interfering magnetic field.

Conductance: A measure of the ability of a battery to carry current. When a difference of electrical potential is placed across a conductor, its movable charges flow, and an electric current (amperes) appears. Conductance is the inverse of impedance.

Controller area network (CAN): A serial bus network of microcontrollers that connects devices, sensors, and actuators in a system or subsystem for real-time control applications. There is no addressing scheme used, as in the sense of conventional addressing in networks (such as Ethernet). Rather, messages are broadcast to all the nodes in the network using an identifier unique to the network. Based on the identifier, the individual nodes decide whether or not to process the message and also determine the priority of the message in terms of competition for bus access. This method allows for uninterrupted transmission when a collision is detected, unlike Ethernets that will stop transmission upon collision detection. Controller area networks can be used as an embedded communication system for microcontrollers as well as an open communication system for intelligent devices.

Counterpoise: A conductor or system of conductors used as a substitute for earth or ground in an antenna system.

Delta-to-Wye transformer: A transformer in which the primary windings are in a Delta configuration and the secondary windings are in a Wye.

Deutsches Institut für Normung (DIN): The German national standards institution.

Differential GPS (DGPS): A system of land-based sites that broadcasts correction signals for improved GPS position accuracy.

Electromagnetic compatibility (EMC): The ability of electronic equipment to operate in proximity with other electrical and electronic equipment without suffering from, or causing, impaired performance.

Electromagnetic interference (EMI): Electromagnetic or electrical disturbances that result in undesirable responses, or impede the performance of other electrical or electronic equipment. For example, EMI is a common cause of deviation in compasses and electronic compass sensors.

Ethernet: Originally developed by Xerox in 1976, it now describes a diverse family of frame-based computer networking technologies for local area networks (LANs). It is the basis for IEEE standard 802.3.

False ground: For the purposes of this book, a ground connection to neutral at a location other than the source of power. In other words, a point where an AC neutral conductor is "falsely" connected to the AC grounding (earth) conductor.

Galvanic isolator: A device installed in series with a boat's earth-ground conductor that blocks low-level DC current (galvanic current) but allows passage of AC fault current.

Ground fault circuit interrupter (GFCI): A device intended to protect people by interrupting an AC circuit whenever its current limit is exceeded.

Ground fault protection (GFP) device: A device intended to protect equipment by interrupting the electric current to the load when a fault current to ground exceeds a predetermined value. Also known as an RCD (residual current device) or an EPD (equipment protecting device).

Harmonic distortion: A load where the wave shape of the steady-state current does not follow the wave shape of the applied voltage. The presence of harmonics that change the AC voltage waveform from a simple sinusoidal waveform to a complex waveform. Harmonic distortion can be generated by a load and fed back into the AC mains, causing power problems to other equipment on the circuit.

Impedance: The opposition to the flow of alternating current in a circuit. It's the high-frequency equivalent of resistance.

Inductance: A property of a conductor or coil that determines how much voltage will be induced in it by a change in current.

Ingress protection (IP): An internationally recognized standard that addresses both water ingress and dust ingress into equipment. The higher the IP number assigned to a given piece of equipment, the more waterproof and dust-proof it is.

Inmarsat: A satellite communications company that provides voice and data services to the maritime, transportation, and aeronautics markets as well as providing general Internet access from any location on the globe. It was founded in 1979 as a British intergovernmental organization (IGO) to serve the maritime industry. It became a private company in 1999 and a public company in 2005.

Institute of Electrical and Electronics Engineers (IEEE): A nonprofit organization that creates internationally recognized standards applicable to electrical and electronic equipment.

International Maritime Organization (IMO): A specialized agency of the United Nations that develops standards for safety at sea, shipping security, and prevention of water pollution as related to ships.

Japanese Industrial Standard (JIS): Specifies the standards used for industrial activities in Japan.

Linear load: In this context, a simple resistive load that draws current in proportion to the voltage delivered. For example, a simple incandescent lightbulb or an electric water heater.

Marine cranking amps (MCA): The number of amps that a new, fully charged battery at 32°F (0°C) can deliver for 30 seconds, and maintain a voltage of 1.2 volts per cell or higher.

Megohmmeter: A specialized ohmmeter for measuring extremely high resistance values, sometimes referred to as an insulation resistance tester. Readings are typically in the millions of ohms, far beyond the reading capabilities of a digital volt-ohmmeter (DVOM). It is often called a "megger."

Milligauss (mG): $\frac{1}{1,000}$ of 1 gauss, which is a unit used for measuring the flux density of magnetic fields. Milligauss are useful for measuring magnetic field levels commonly found in the environment.

Multiplexer: A communications device (black box) that combines several signals for transmission over a single medium.

National Electric Code (NEC): The preeminent set of safety standards for electrical installations. Developed and published by the National Fire Protection Association (NFPA).

National Marine Electronics Association (NMEA): The foremost standards group in North America that deals with marine electronics installations.

Network: A group of two or more computer systems linked together to share information and hardware.

Network architecture: The structure of a communications network. An *open architecture* allows adding, upgrading, and swapping of components. It can be connected easily to devices and programs made by other manufacturers. It uses off-the-shelf components and conforms to approved standards. A *closed architecture* has a proprietary design, making it difficult to connect the system to other systems. The hardware manufacturer chooses the components, and they are generally not upgradable.

Nonlinear load: In contrast to a linear load, a load that will not have constant values in current, voltage, or resistance. Examples are electronic control circuits for refrigeration systems and motor loads.

Peak capture: A feature found on better electrical measuring instruments that allows the user to capture and record the maximum value measured on a circuit with fluctuating values.

Power factor: The ratio of real power to apparent power in an AC system. The amount of current and voltage the customer actually uses compared to what the utility supplies.

Protocol: In the context of data communication, a common set of rules, signals, and data structures (for either hardware or software) that governs how computers and other network devices exchange information over a network.

Radiated power: The product of antenna input power and antenna power gain, expressed in kilowatts.

Radio frequency interference (RFI): Radio signals normally emitted from an electrical device that negatively impact the quality of reception in another electronic or electrical device. Symptoms vary and can include loss of reception or, in a television, reduced picture quality. RFI is a common problem on boats and a primary source of equipment performance problems.

Reactive power: The difference between the electrical power delivered to a system and the power converted to useful work. Reactive power is stored energy that is returned to the power source.

Real power: The amount of electrical energy that is converted into useful work. Also known as active power or working power.

Residual current: In this context, leakage current in an AC circuit. In an AC circuit, the hot and neutral conductors cancel each other out in a true-sine-wave power supply system. When a current differential occurs between the neutral and hot, this measured difference is referred to as residual current, and it is leaking somewhere into the grounding system.

Sealed valve regulated (SVR) batteries: A pressurized lead-acid battery that is completely sealed, and has a pressure-controlled vent to release excess pressure inside the battery case (e.g., due to overcharging).

Sentence structure: In NMEA parlance, describes the sequence and type of data information codes that are distributed throughout the network. (For specific sentence codes, see the sidebar on page 151.)

Sinusoidal: Having a succession of waves or curves, as in an AC sine wave.

Specific gravity: The ratio of the density of a material to the density of water. In a lead-acid battery, the specific gravity of the electrolyte is a measure of the battery's state of charge.

Standing wave ratio (SWR): The ratio of the maximum RF current to the minimum RF current on the line. It can be thought of as a measure of an antenna system's efficiency.

Time domain reflectometer (TDR): A tool used to measure time delays in signal transmission through conductors that correlate to excessive resistance, or open or shorted circuits.

True root mean square (RMS): In an electrical context, describes the algorithm a meter must have to accurately analyze a less than perfect AC waveform.

Uninterruptible power supply (UPS): A small battery that provides temporary power to electronic equipment when the connected power supply experiences an extreme voltage drop or total loss.

Velocity of propagation (VOP): In an electrical context, a percentage of the speed of light that describes the speed of current through a cable. Different conductors have different VOP values due to their inherent resistances to electrical current flow.

ABYC	American Boat and Yacht Council	**IEEE**	Institute of Electrical and Electronics Engineers
AFCI	arc fault circuit interrupter	**IMO**	International Maritime Organization
AGM	absorbed glass mat		
AIC	ampere interrupting capacity	**IP**	ingress protection
ASCC	available short-circuit current	**JIS**	Japanese Industrial Standard
AWG	American Wire Gauge	**kW**	kilowatt
CAN	controller area network	**mA**	milliamp
CCA	cold cranking amps	**MARPA**	mini automatic radar plotting aid
dB	decibel		
DGPS	Differential GPS	**MCA**	marine cranking amps
DIN	Deutsches Institut für Normung	**mG**	milligauss
DVOM	digital volt-ohmmeter	**MHz**	megahertz
EMC	electromagnetic compatability	**NEC**	National Electric Code
EMI	electromagnetic interference	**NMEA**	National Marine Electronics Association
EPD	equipment protecting device		
FCC	Federal Communications Commission	**PC**	personal computer
		RCD	residual current device
GFCI	ground fault circuit interrupter	**RFI**	radio frequency interference
GFP	ground fault protection	**RMS**	root mean square
GHz	gigahertz	**SWR**	standing wave ratio
GMDSS	Global Maritime Distress and Safety System	**TDR**	time domain reflectometer
		UPS	uninterruptible power supply
GPS	global positioning system	**VAC**	volts of alternating current
Hz	Hertz	**VDC**	volts of direct current
IEC	International Electrotechnical Commission	**VOP**	velocity of propagation
		WAAS	wide area augmentation system

INDEX

E P B M We hope you enjoyed this title
from Echo Point Books & Media

Before Closing this Book, Two Good Things to Know

Buy Direct & Save

Go to www.echopointbooks.com (click "Our Titles" at top or click "For Echo Point Publishing" in the middle) to see our complete list of titles. We publish books on a wide variety of topics—from spirituality to auto repair.

Buy direct and save 10% at www.echopointbooks.com

DISCOUNT CODE: EPBUYER

Make Literary History and Earn $100 Plus Other Goodies Simply for Your Book Recommendation!

At Echo Point Books & Media we specialize in republishing out-of-print books that are united by one essential ingredient: high quality. Do you know of any great books that are no longer actively published? If so, please let us know. If we end up publishing your recommendation, you'll be adding a wee bit to literary culture and a bunch to our publishing efforts.

Here is how we will thank you:

- A free copy of the new version of your beloved book that includes acknowledgement of your skill as a sharp book scout.
- A free copy of another Echo Point title you like from echopointbooks.com.
- And, oh yes, we'll also send you a check for $100.

Since we publish an eclectic list of titles, we're interested in a wide range of books. So please don't be shy if you have obscure tastes or like books with a practical focus. To get a sense of what kind of books we publish, visit us at www.echopointbooks.com.

If you have a book that you think will work for us,
send us an email at editorial@echopointbooks.com

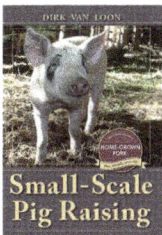

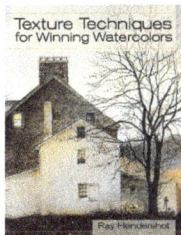

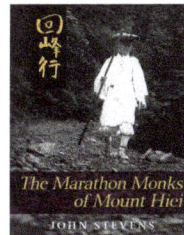

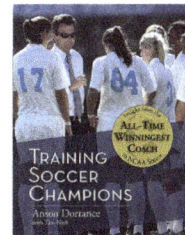